STUDIES IN MATHEMATICAL BIOLOGY

PART I: CELLULAR BEHAVIOR AND THE DEVELOPMENT OF PATTERN

MAA STUDIES IN MATHEMATICS

Published by

THE MATHEMATICAL ASSOCIATION OF AMERICA

———

Committee on Publications
E. F. BECKENBACH, Chairman

Subcommittee on MAA Studies in Mathematics
G. L. WEISS, Chairman
F. J. ALMGREN, JR.
WANDA W. HELM
A. C. TUCKER

Studies in Mathematics

The Mathematical Association of America

Michael A. Arbib
University of Massachusetts

Jack D. Cowan
University of Chicago

G. B. Ermentrout
University of Chicago

Warren J. Ewens
University of Pennsylvania

John M. Guckenheimer
University of California,
Santa Cruz

Samuel Karlin
Stanford University

Stuart Kauffman
University of Pennsylvania

Nancy Kopell
Northeastern University

Simon A. Levin
Cornell University

Donald Ludwig
University of British Columbia

Robert H. MacArthur
(deceased)

Robert M. May
Princeton University

Thomas Nagylaki
University of Chicago

George F. Oster
University of California, Berkeley

John Rinzel
National Institutes of Health,
Bethesda

S. I. Rubinow
Cornell University Medical School

Lee A. Segel
Rensselaer Polytechnic Institute
and
Weizmann Institute of Science,
Rehovot, Israel

Arthur T. Winfree
Purdue University

E. C. Zeeman
University of Warwick,
Coventry, England

Studies in Mathematics

Volume 15

STUDIES IN MATHEMATICAL BIOLOGY
PART I: CELLULAR BEHAVIOR AND THE
DEVELOPMENT OF PATTERN

S. A. Levin, editor
Cornell University

Published and distributed by
The Mathematical Association of America

©1978 by
The Mathematical Association of America (Incorporated)
Library of Congress Catalog Card Number 78-53425

Complete Set ISBN 0-88385-100-8
Vol. 15 ISBN 0-88385-115-6

Printed in the United States of America

Current printing (last digit):

10 9 8 7 6 5 4 3 2 1

ACKNOWLEDGEMENTS

Professor Guckenheimer's contribution, "Isochrons and Phaseless Sets," is reprinted with permission of Springer-Verlag, New York, from the *Journal of Mathematical Biology*, 1 (1975), pages 259-273; Professor Kauffman's paper, "The Large Scale Structure and Dynamics of Gene Control Circuits: An Ensemble Approach," is reprinted from the *Journal of Theoretical Biology*, 44 (1974), pages 167–190, by permission of Academic Press, New York; Appendix 3 of Professor May's article, "Mathematical Aspects of the Dynamics of Animal Populations," is a reprint of "Niche Overlap as a Function of Environmental Variability" by Robert M. May and Robert H. MacArthur, which first appeared in *The Proceedings of the National Academy of Sciences U.S.A.*, 69 (1972), pages 1109–1113; the figures in Professor Oster's article, "The Dynamics of Nonlinear Models with Age Structure," are reprinted with permission of Springer-Verlag, New York, from "Dynamics of density dependent populations models", an article which appeared in the *Journal of Mathematical Biology*, Volume 4, No. 2, and was co-authored by Professor Oster with Professors Guckenheimer and Ipaktchi; a portion (pages 158–168) of Professor Segel's contribution, "Mathematical Models for Cellular Behavior," is reprinted from "Mathematics Applied to Deterministic Problems in the Natural Sciences" (pages 22–30) by Lee Segel and C. C. Lin, with permission of Macmillan Publishing Co., Inc., New York; Professor Winfree's article, "Patterns of Phase Compromise in Biological Cycles," is reprinted from the *Journal of Mathematical Biology*, 1 (1974), pages 73–95, with permission of Springer-Verlag, New York; Professor Zeeman's paper, "Primary and Secondary Waves in Developmental Biology," is reprinted with permission of the publisher, American Mathematical Society, from *Lectures on Mathematics in the Life Sciences*, Copyright © 1974, Volume VII, pages 69–96.

INTRODUCTION

Ideally, any biologist's province includes the whole of biology; as distant as phenomena at the molecular and the community or ecosystem levels may seem from one another, they are ultimately linked in effect and in evolution. In practice, however, the nature of research usually forces a narrower focus, restricted with regard to level of organization.

Thus, what has made mathematical biology a somewhat coherent discipline has been less a unity of biological purpose than a similarity in the formal natures of the problems faced, and the kinds of questions asked. How are systems organized dynamically? What are the relationships between the behaviors of single units and of large-scale ensembles of such units? How does one recognize and understand pattern at the myriad structural, spatial, and temporal scales on which it is expressed?

These questions are of more general interest than the specific biological problems which engender them, and distinct biological problems often lead to similar mathematical formulations. Questions of pattern formation, for example, have arisen in the study of neural, developmental, genetical, and ecological systems, as well as in totally non-biological contexts; these have in many instances found common expression as variants of diffusion-reaction schemes. The analysis of these and other such mathematical abstractions has not only provided new directions for mathematical research; it has also contributed significantly to biological thought.

This set of two volumes on mathematical problems in theoretical biology is more than simply a collection of unrelated papers. The commonalities and interfaces between the various problems and approaches forcefully demonstrate the maturity of mathematical biology as a discipline; that this maturity has evolved largely from recent work is testimony to its dynamic and vibrant nature. These studies cannot be comprehensive; missing, for example, are such topics as immunology, biomechanics, and biostatistics. Nonetheless, it is hoped that by reading through them, anyone will find places where he or she can contribute, and without major scientific plastic surgery.

Volume 15 basically examines questions at the organismic and suborganismic levels, while Volume 16 is directed towards population phenomena. However, it is clear that many questions at the organismic and suborganismic levels are basically population problems. Indeed, at least one paper in the second volume (that by Sol Rubinow) is motivated by suborganismic phenomena.

Volume 15 opens with three papers devoted to an understanding of nervous system function. John Rinzel, beginning from the Nobel Prize winning work of Hodgkin and Huxley, develops the theory of signal transmission by single nerve cells, surveying past and current theories and examining in detail what is known regarding solutions. He proceeds to the next level, dendritic integration, a theory which has been largely developed by W. Rall and Rinzel.

Rinzel's paper is a natural introduction to that of Cowan and Ermentrout, which is concerned primarily with populations of neurons and their aggregate behavior. As Arbib suggests in the third paper, this is basically a "bottom-up" approach to the brain; it is neatly complemented by Arbib's own approach to brain theory, which derives inspiration from the successes in the field of artificial intelligence. This group of papers draws on a diverse spectrum of mathematical methods, including the diffusion-reaction approach and bifurcation and catastrophe theory, and demonstrates a range from empiricism to speculation that will be characteristic throughout these volumes.

These first three papers provide an example of one complex system with unique characteristics at different levels of organization; Lee Segel follows with a lucid discussion of the emergence of "collective" behavior in a variety of cellular systems. Nancy Kopell then concludes the middle section with a most effective survey of pattern formation in the reaction-diffusion systems which are so pervasive in these Studies.

The volume concludes with a sequence of reprints addressed to general problems of development and of biological rhythms. Christopher Zeeman, using the formalism of Thom's catastrophe theory, speculates on possible applications to embryology. Catastrophe theory, a subject which has generated remarkable debate in recent years, also was involved earlier in this volume in the contribution of Cowan and Ermentrout. This reprint of Zeeman's writings is a selection from a much larger paper, "Primary and Secondary Waves in Developmental Biology" (pages 69–161 in S. A. Levin, ed., 1974, *Lectures on Mathematics in the Life Sciences, Volume 7: Some Mathematical Questions in Biology VI*), and is reprinted by permission of the American Mathematical Society. The very section which is reprinted here has come under severe attack recently (R. S. Zahler and H. J. Sussman, 1977, "Claims and accomplishments of applied catastrophe theory," *Nature*, 269:759-763); a balanced view of the controversy is provided by John Guckenheimer (*Mathematical Intelligencer*, 1:1, 1978).

Stuart Kauffman's paper, "The Large Scale Structure and Dynamics of Gene Control Circuits: An Ensemble Approach," devoted to an understanding of gene control systems, is reprinted by permission of the *Journal of Theoretical Biology* (Vol. 44 (1974):167–190).

In the last pair of papers, Arthur Winfree and John Guckenheimer develop mathematical models for phase-resetting experiments on biological clocks, state a number of related mathematical questions, and provide answers to some of those questions. These two papers, both reprinted from Volume 1

(1974–1975) of the *Journal of Mathematical Biology* (pages 73–95 and 259–273), provide an excellent example of the interplay between experimentalists and mathematicians in the statement and solution of problems of both biological and mathematical significance.

Much is left out of these volumes, and there are a wide variety of other sources available to the reader who seeks to go beyond what is to be found here. In addition to numerous more specialized journals, the *Journal of Theoretical Biology*, the *Journal of Mathematical Biology*, *Mathematical Biosciences*, the *SIAM Journal on Applied Mathematics*, and the *Bulletin of Mathematical Biology* are broad-spectrum journals publishing papers in the applications of mathematics in biology. More expository articles are to be found in the series *Lectures on Mathematics in the Life Sciences*, published by the American Mathematical Society for over a decade, and monographic contributions in the series *Lecture Notes in Biomathematics*, published by Springer-Verlag. Any bibliography attempted at this point would become quickly out of date; however, it is hoped that the articles in these volumes will provide a sufficient entrée into a vast and rapidly expanding literature.

It has taken a long time to create this collection, and some of the original contributions were written in 1975 or 1976. The delay in publication may render some aspects of the older articles out-of-date, and for this I accept full responsibility.

SIMON A. LEVIN

Ithaca, New York
July 28, 1976

CONTENTS

INTEGRATION AND PROPAGATION OF NEUROELECTRIC SIGNALS

John Rinzel

1. INTRODUCTION

A primary form of communication in and through the nervous system is the transmission of electrical signals by nerve cells or neurons. Sensory input is transduced to electrical activity. Muscular contraction is controlled by neural signaling. Higher thought processes reflect patterns of activity in complex neural pathways. At various levels of neural processing an individual cell may be called upon to combine or *integrate* the input signals it receives from many other neurons. In response, it develops a pattern of activity which it may *propagate* to distant nerve or muscle cells over its long nerve fiber. Here we will describe mathematical models for these two aspects of neurocommunication at the cellular level. The treatments are basically continuous in nature. They involve ordinary and parabolic partial differential equations to describe the spatio-temporal patterns of voltage and current flow throughout an individual neuron.

These models for single cell activity were, and continue to be, formulated largely by physiologists and biophysicists. They are based upon electrical and anatomical data from a number of experimentally convenient preparations. The 1952 landmark work of Hodgkin and Huxley [40] illustrates the impressive degree of quantitative accountability which has been achieved in a few cases over the past twenty-five years. While sufficient experimental data are not yet available to determine full quantitative descriptions for the behavior of all classes of neurons, the qualitative features of the present models are reasonably assumed to be characteristic of many types of cells. At both the quantitative and qualitative levels, insight obtained from the models can aid the physiologist in the interpretation and evaluation of electrophysiological data and techniques. Beyond this, functional implications and mechanisms for various signaling phenomena have been proposed or deduced as a consequence of the modeling. Such direct applications offer the opportunity for mathematical contributions to neuroscience. In addition, the rich mathematical structure of various models provokes stimulating research questions for the mathematician.

This chapter will not touch on all the interesting physiological and mathematical aspects of cellular signaling. Nor will it attempt to provide an in-depth historical account or complete reference list. Rather, it will outline a few currently used models with some applications and analyses for them. The following section offers a more technical introduction to nerve signaling and previews the subsequent sections. Additional introductory sources to neurophysiology are Stevens [80], Hodgkin [39], and Katz [48]. Among the interesting surveys which discuss mathematical models for neuroelectric signaling are the extensive articles by FitzHugh [29], Noble [55], Rall [66], Scott [77], and the books by Cole [11], Jack, Noble, and Tsien [47]. Portions of this chapter have appeared elsewhere [72, 73].

2. PHYSIOLOGY, PHENOMENOLOGY, AND PHYSICAL MODELS

To introduce the terminology and signaling phenomenology of cellular neurophysiology I refer to the schematic neuron in Figure 1. It represents a motoneuron from the spinal cord and its primary function is to control muscular contraction. Three basic structural components may be identified. Extensively branched tree-like extensions called *dendrites* emanate from the cell body or *soma*. The scale and neuron in Figure 1 indicate typical dimensions for dendritic lengths and soma size for a variety of neurons. Dendritic branch diameters may measure tens of microns for the trunks and decrease down to microns for the fine terminal branches. The nerve fiber or *axon*, which is usually long relative to the dendrites, also originates at the soma; the axon in Figure 1 has arrows positioned alongside it. Axon lengths vary over the range of millimeters to meters and their diameters are on the order of microns to tens of microns. For many nerve cells, the axon is relatively unbranched except near its terminal end where profuse arborization may be seen.

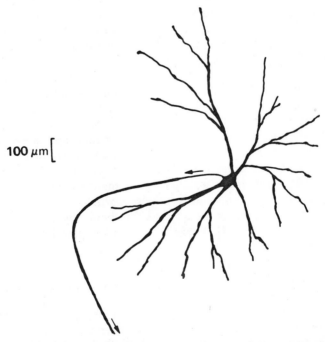

100 μm

FIG. 1. Schematic motoneuron illustrating soma, branching dendritic trees, and part of axon. (Redrawn from Shepherd [78] with permission of Oxford University Press.)

The entire neuron is encased by a thin membrane which is permeable to varying degrees to different ions in the physiological environment. Its properties are typically not uniform over the neuron surface. The complex biophysical nature of nerve membrane permits a broad range of signaling phenomenology. Membrane potential, the voltage difference across the nerve membrane, is the primary observable of neuroelectric activity; it is measured in millivolts (mV). The currents which flow intracellularly and across the membrane are ionic. Spatio-temporal distributions of current and membrane potential are the signals by which cellular communication is achieved.

Such variations in structure, as geometrical arrangement and membrane properties, between neuron components are frequently correlated with differences in signaling function. First to discuss axonal signaling we introduce the notion of the nerve impulse. It may loosely be described as the fundamental unit of long range nervous communication. Somewhat more precisely, it refers to a brief, transient chemical-electrical event which travels as a wave along the nerve fiber; a typical time scale is milliseconds (msec). The membrane potential transient of the impulse is called the action potential; its peak value is around 100 mV above the resting potential.

For a long, uniform axon propagating a single impulse, the propagation speed is constant, on the order of 1–100 meters per second and the nerve is at rest sufficiently far in front of and behind the wave. Impulse initiation requires a sufficiently strong stimulus. A weak stimulus produces only a locally observable response. Hence, one says there is a threshold for nerve excitation. The characteristics of a steadily propagating impulse, such as speed and action potential shape, are unique to the nerve. They do not vary with the strength of an adequate stimulus. Because of this property, physiologists describe the impulse as an "all or none" response. If brief stimuli are applied at some fixed location, a cell body for example, and successively in time they may initiate successive impulses provided the stimuli are strong enough and sufficiently separated in time. The minimum separation time for two stimuli to be successful is referred to as the absolute refractory period and the threshold for the second impulse initiation depends upon the separation time. If a strong stimulus is steadily maintained the nerve may respond with an indefinite train of identical impulses. The frequency for such repetitive firing depends on the stimulus strength.

Just these few qualitative features of impulse propagation and excitation illustrate the high degree of nonlinearity involved. Biological and chemical systems with dynamic properties similar to these have been described as "excitable" and nerve axon is often cited as a familiar example, e.g., [82]. In terms of physiological function, the axon's primary role, nerve conduction, is the transmission of sequences of nerve impulses from the cell body to other distant neurons or muscle fibers. Information is thought to be coded as some functional of the pulse frequency. In some contexts it may be instantaneous frequency; in others, perhaps some time average. The axon is thus usually considered as the neuron's output limb and the arrows in Figure 1 indicate the usual direction of impulse propagation.

Input from the axons of other neurons and sensory cells is delivered primarily to a cell's dendritic trees and soma. The discrete input locations are called synapses and they are distributed mostly over the many dendritic branches. There is generally a large number of synapses on a cell; perhaps, tens of thousands. The distribution of synapses and the spatio-temporal pattern of their activation determines the voltage and current supplied to the cell body and axon base, and consequently determines the cell's impulse firing pattern. A membrane potential change generated by a synaptic input is called a post-synaptic potential (PSP). The dendrites provide convergent pathways for transmission of PSP's to the cell body. In various types of cells PSP's are thought to suffer attenuation during dendritic transmission. Individually generated PSP's are also thought to summate linearly. These qualitative features may be contrasted with unattenuated conduction and nonlinear interaction of action potentials in axons. Neuronal integration refers to the manner in which synaptic input signals are transmitted and combined in the dendritic trees and cell body.

The above description of cellular structure and function can be considered classical. However, because there are many different types of neurons, it is not a universal description [78]. Some cells have no axons, others are without dendrites. A few types of axon apparently do not conduct impulses [78] while, according to some evidence, certain dendritic systems may propagate action potential-like signals [50]. In addition to axodendritic synapses one finds dendro-dendritic synapses in some populations; these apparently mediate both input and output signaling at the dendritic level [63]. While the mathematical models we will discuss could incorporate these various features we will introduce them with the classical description in mind. Yet because the models are adaptable, they provide quantitative tools for examining various hypotheses for neuronal function and mechanism. The various possibilities pose interesting and exciting research questions.

We begin by considering an unbranched dendritic or axonal segment with uniform circular cross-section of diameter d (cm). For the "core conductor" physical model, the segment is treated as a tube of thin membrane which surrounds the intracellular conducting gel. A constant ohmic resistivity R_i (ohm cm) characterizes the intracellular core. Per unit area, the membrane is electrically represented as a constant capacitance C_m ($\mu F/cm^2$) in parallel with conducting pathways for the various ionic currents and, for the dendritic case, any synaptic current. For determining membrane potential and currents, the extracellular resistivity is usually neglected so that the extracellular medium is effectively isopotential. Typically, the only significant spatial variable is axial distance x along the segment. This specialization of the core conductor leads to one dimensional cable theory and an electrical transmission line as a physical model.

Let $v(x, t)$ denote the deviation of membrane potential from its constant resting value. When $v(x, t) > 0$ the membrane is said to be depolarized. From the above physical description, the local membrane current density (per unit area) is

$$I_m = C_m \frac{\partial v}{\partial t} + I_{ion} + I_{syn}.$$

The first term represents the capacitive component. I_{ion} corresponds to the combination of individual ionic currents which depend on $v(x, t)$ and the membrane's inherent permeabilities for the various ion species. I_{syn} is the current generated by synaptic sources; it also is ionic in nature. However, because synapses are typically not locally self-activated we distinguish I_{syn} from the autonomous membrane ionic currents I_{ion}. The standard convention has outward membrane current as positive. According to Ohm's law the axial or intracellular core current i_i, taken positive with increasing x, is given by

$$i_i = -\left(\pi d^2/4R_i\right)\partial v/\partial x. \tag{1}$$

Conservation of current requires that the membrane current per unit length $I_m \pi d$ equals the negative gradient or loss per unit length of axial current $-\partial i_i/\partial x$ plus $I_{app}\pi d$, the current per unit length applied by an experimenter's electrode. From this balance relationship, follows a parabolic equation for $v(x, t)$

$$C_m \frac{\partial v}{\partial t} = \frac{d}{4R_i} \frac{\partial^2 v}{\partial x^2} - I_{ion} - I_{syn} + I_{app}. \tag{2}$$

To interpret our sign conventions, consider (2) with $\partial v/\partial x = 0$ and $I_{app} = 0$. Then if $v > 0$, outward current, $I_{ion} + I_{syn} > 0$, tends to decrease v back toward rest while inward current increases v and further depolarizes the membrane from rest.

For axonal membrane, we focus on excitability as the distinguishing feature and set $I_{syn} = 0$. The mechanism for nerve conduction originates with the intrinsic ionic currents which depend nonlinearly upon $v(x, t)$ and autonomously on time. Auxiliary membrane variables $w_i(x, t)$, $1 \leqslant i \leqslant n$, are introduced to describe the currents and conductances for the individual ionic species. Their evolution depends only upon the membrane state at x, described by $v(x, t)$ and $w_i(x, t)$, and not directly upon axial gradients. They satisfy autonomous ordinary differential equations in time. Hence, for dimensionless variables (see [30] for scaling) the structural form of a nerve conduction equation is a nonlinear parabolic system of partial differential equations:

$$\frac{\partial \mathbf{V}}{\partial t} = \begin{bmatrix} \dfrac{\partial^2 v}{\partial x^2} \\ 0 \\ \vdots \\ 0 \end{bmatrix} + \mathbf{F}(\mathbf{V}), \tag{3}$$

where $\mathbf{V}$ is the $(n + 1)$-vector $(v, w_1, \ldots, w_n)$ and $\mathbf{F}$ is an $(n + 1)$-vector function of $\mathbf{V}$. In this autonomous form no external inputs are represented and the membrane characteristics are assumed homogeneous along the fiber. The typical theoretical axon is unbranched and usually infinite or semi-infinite in length.

In section 3.1 we will introduce the empirical equation of Hodgkin and Huxley for squid axon; for it, $n = 3$. They, as have other investigators for different preparations, determined $\mathbf{F(V)}$ from electrical data for a dissected and extracted section of axon. From such data we will motivate, in section 3.2, the analytically more convenient FitzHugh-Nagumo equation. It mimics the essential qualitative phenomena and has only two variables, v and one auxiliary variable.

For appropriate parameter values these equations have traveling wave solutions. They correspond to the solitary nerve impulse and to periodic trains of impulses for repetitive firing. These particular solutions satisfy nonlinear ordinary differential equations. As solutions to the partial differential equations, some of the waves are stable and others unstable. For a piecewise linear FitzHugh-Nagumo equation, this structure can be studied analytically in detail. For other equations, these statements are based partly upon analytic but primarily upon numerical results. In section 3.3 we will exploit the simple model equation to describe the traveling waves, introduce appropriate notions of stability, and discuss the waves' stability.

Recall that nerve impulses are generated in response to particular kinds of stimuli described as suprathreshold. Analogously for the mathematical models, traveling waves are asymptotic solutions to correctly posed initial-boundary value problems for (3) with appropriate data. In section 3.4 we will formulate some of these problems, describe quantitative notions of threshold, and discuss some stimulus-response properties of the models. We will also briefly describe some interesting problems which arise from structural variations in axons.

For dendritic membrane, the ionic currents are generally thought to be linear functions of membrane potential over a larger range of potential than for axonal membrane. In other words, the voltage threshold for excitation is assumed to be reasonably high.

For this reason, models for dendritic integration (e.g., for the motoneuron) usually assume that $I_{ion}(x, t)$ may be approximated by $R_m^{-1}v(x, t)$ where R_m is the constant membrane resistance. Such a model describes passive membrane properties. The current I_{syn} generated by active synapses is also usually modeled as a linear function of membrane potential. Hence equation (2) for $v(x, t)$ in a single dendritic branch is a linear parabolic partial differential equation. The qualitative features of its solutions are known. For such an unbranched dendritic segment, the equation can, in some cases, be solved explicitly with classical techniques. The principal mathematical complication is due to the extensively branched geometry.

In section 4.1 a model for dendritic integration will be formulated in greater detail to treat the many branches and the cell body. Next, in section 4.2, I will describe an idealized branching geometry, first introduced by Rall [60], which considerably enhances analytical tractibility. With this simplification, Rall's model has been used to qualitatively understand the functional importance of dendritic synaptic activity. It has also been applied quantitatively. Mathematical solutions account for experimentally recorded potentials in certain classes of neurons and provide recipes for estimating neuronal electrical parameters. I will describe these applications in section 4.3 in addition to analytic results which enable theoretical comparisons for effectiveness of input delivered to individual synaptic locations. I will also illustrate two applications of coupled dendritic-axon models to simulate various aspects of whole neuron behavior.

3. AXONAL IMPULSE PROPAGATION

3.1. The Hodgkin-Huxley equation. The most well-known and widely accepted quantitative description of ionic current flows for axonal membrane is that of A. L. Hodgkin and A. F. Huxley [40]. Here we will outline their elegant synthesis of a model. It is based upon data obtained from the giant axon of squid. Because of its large diameter, on the order of .5 mm, this fiber is experimentally convenient. To construct their model the investigators, of course, had to measure the membrane ionic current I_{ion}. According to (2) with $I_{syn} = 0$, I_{ion} is isolated and equal to the applied current I_{app} when the capacitive and intracellular axial currents are eliminated. This observation was implemented through an ingenious experimental technique, the voltage clamp.

For a typical experiment, a length of axon is removed from the squid and a thin wire electrode is threaded lengthwise along the core. Because the electrode has low resistivity relative to the core, spatial variations of current and potential are virtually eliminated. For this "space clamped" region (a few cm. in length), $\partial v / \partial x = 0$ in (2). Furthermore, with appropriate electronics, one can hold the membrane potential of the space clamped axon at any desired constant level. The "voltage clamp" thus eliminates the capacitive current since $\partial v / \partial t = 0$. When the clamp is turned on at $t = 0$, the ionic conductances begin changing with time and the control circuit must apply the appropriate current I_{app} through the intracellular wire to maintain v at the specified value. In this space and voltage clamped configuration we use the notation $I_{ion}(x, t) = I_{ion}(t; v)$ and as (2) implies, $I_{ion}(t; v) = I_{app}(t; v)$.

A few such current profiles $I_{ion}(t; v)$ are shown in Figure 2 for selected values of v. These are actually theoretical curves calculated from the Hodgkin-Huxley (HH) equation (6). They are essentially identical to the experimental ones. Qualitatively, the curves for $v < 120$ mV indicate an early transient inward flow of ions, $I_{ion} < 0$, followed by a late steady outward flow, $I_{ion} > 0$.

Such information by itself however does not lead to an insightful theory. The major HH contributions were the separation of I_{ion} into its individual ionic currents (primarily for sodium Na^+ and potassium K^+ ions), demonstration of their independence, and derivation of empirical expressions for them and the ion conductances. To do this they hypothesized the form of I_{ion} as

$$I_{ion} = g_{Na}(v - \bar{v}_{Na}) + g_K(v - \bar{v}_K) + \bar{g}_L(v - \bar{v}_L). \qquad (4)$$

This represents the individual components as independent and phenomenological ohmic currents. Each is the product of a conductance and a potential difference. In (4), g_{Na} and g_K are voltage and time dependent membrane conductances for Na^+ and K^+

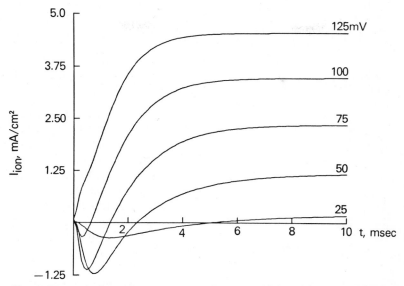

FIG. 2. Theoretical curves of voltage and space-clamped current, $I_{ion}(t; v)$ versus t, computed for the HH model. $I_{ion}(t; v)$ is evaluated, for constant v, from equations (4), (5), and the last three of (6).

ions; they are positive valued. The quantities $\bar{v}_{Na}$ and $\bar{v}_K$ are constant (Nernst) potentials which are determined by the ionic concentrations inside and outside the nerve. In normal physiological solutions, $\bar{v}_{Na} \doteq 115$ mV and $\bar{v}_K \doteq -12$ mV. Thus, for $v < 115$ mV, sodium current is inward and for $v > -12$ mV the potassium current is outward. The last term in (4) corresponds to a functionally less significant and relatively small amplitude current due to leakage of other ions; for it, $\bar{g}_L$ and $\bar{v}_L$ are constant.

Because $\bar{v}_{Na}$ and $\bar{v}_L$ depend on the environmental ionic concentrations these potentials can be adjusted by altering those concentrations. Hodgkin and Huxley exploited this fact to separate the individual currents in (4). By altering $\bar{v}_{Na}$ and then clamping v to the altered $\bar{v}_{Na}$, the sodium current could be eliminated. From the potassium current and its ohmic representation, the potassium conductance $g_K(t; v)$ was determined for various values of the clamping potential v. Then from the current records under normal conditions (Figure 2) one could determine, by subtraction, the sodium current and $g_{Na}(t; v)$.

During voltage clamp $g_{Na}(t; v)$ rapidly turns on and then more slowly turns off again. Two variables, m and h, were introduced to model this behavior; they are referred to as sodium activation and sodium inactivation. The potassium conductance has only a slow turn-on characteristic which was represented in terms of potassium activation, n. With remarkable persistence, Hodgkin and Huxley could accurately fit the conductance dynamics for the different values of v by using the expressions

$$g_{Na} = \bar{g}_{Na} m^3 h, \quad g_K = \bar{g}_K n^4, \tag{5}$$

where $\bar{g}_{Na}$, $\bar{g}_K$ are constants and by having each of the phenomenological variables (m, h, n) satisfy an equation of the form

$$\frac{\partial \alpha}{\partial t} = \frac{\alpha_\infty - \alpha}{\tau_\alpha},$$

where the "steady state value" α_∞ and the "time constant" τ_α depend on v.

Finally then, the full (unclamped) HH system follows from (2) with $I_{syn} = I_{app} = 0$, and from (4), (5), and the m, h, n evolution equations:

$$C_m \frac{\partial v}{\partial t} = \frac{d}{4R_i} \frac{\partial^2 v}{\partial x^2} - \bar{g}_{Na} m^3 h (v - \bar{v}_{Na})$$

$$- \bar{g}_K n^4 (v - \bar{v}_K) - \bar{g}_L (v - \bar{v}_L),$$

$$\frac{\partial m}{\partial t} = (m_\infty(v) - m)/\tau_m(v), \tag{6}$$

$$\frac{\partial h}{\partial t} = (h_\infty(v) - h)/\tau_h(v),$$

$$\frac{\partial n}{\partial t} = (n_\infty(v) - n)/\tau_n(v).$$

The voltage dependent time constant and steady state functions are plotted in Figure 3. Values of the auxiliary variables range

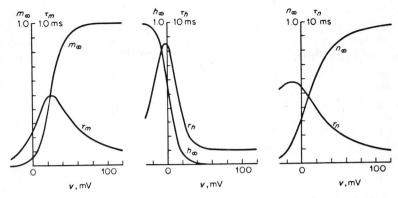

FIG. 3. Curves of the steady-state values and time constants of m, h, and n, as functions of v for the HH model. (As redrawn by FitzHugh [29] from K. S. Cole, *Biophys. J.*, **2** (No. 2, part 2) (1962) 101–119.)

from 0 to 1. As v increases both m_∞ and n_∞ tend to one because m and n are turn-on variables while h_∞ tends to zero since h is a turn-off variable. Also note that the scale for τ_h and τ_n is on the order of 10 msec while for τ_m it is on the order of msec. This is because Na^+ inactivation and K^+ activation are slow "processes" and Na^+ activation is relatively fast. The time constant functions as shown here are for 6.3°C. For any other temperature T, they are divided by $\phi = \exp[0.1(\log 3)(T-6.3)]$.

Even though the HH model is based upon a restricted set of data (voltage clamp) and for a single preparation, its qualitative features are consistent with the classical signaling phenomena described in section 2. Among these features are propagation of a single impulse and trains of impulses, threshold properties for their initiation, appropriate dependence of propagation characteristics and thresholds, upon temperature and other parameters, also subthreshold behavior with a linear regime for small signals. These characteristics were demonstrated primarily by numerical calculation. Some will be discussed further in subsequent sections.

For squid axon, the model quantitatively accounts for many, but not all, experimental observations. As an illustration, Figure 4

compares the HH action potential with an experimentally recorded one for T = 18.5°C. The shape is reasonably matched as
well as the speed, 18.8 m/sec theoretical versus 21.2 m/sec experimental. From the action potential and waveforms for *m*, *h*, and *n*,
theoretical ionic currents may be calculated and used to quantitatively describe the events during the impulse. The membrane
potential at a location in front of the impulse is brought toward
threshold by axial current supplied from the approaching action
potential. This turns on the rapid inward sodium current which
further increases *v*. As *v* approaches 100 mV the sodium current
starts turning off and potassium current starts flowing outward.
This tends to lower *v* and bring it back toward rest.

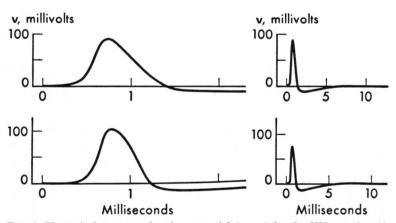

FIG. 4. Theoretical propagated action potential (upper) for the HH equation (6)
and experimentally recorded action potential (lower). Comparisons are shown for
two time scales (left and right). (Redrawn from Hodgkin and Huxley [40].)

The HH contribution is impressive in terms of both model
construction and quantitative accountability. Through it most
physiologists have been exposed at some level to mathematical
modeling. The basic concepts have been applied with quantitative
modifications, for example in the steady state values, rates, or
powers of the auxiliary variables, to describe other excitable nerve

and muscle behavior (e.g., see [55]). Given such broad acceptance for the theory one should not overlook the fact that its primary justification is empirical. The auxiliary variables are not observables nor do they have a rigorous physical interpretation in terms of membrane molecular dynamics. Our understanding of membrane behavior at the molecular level is still incomplete; see [11, 47, 49] for summaries of some current theories. It will be interesting to follow the interpretation given to the HH theory as our understanding develops.

3.2. The FitzHugh-Nagumo equation.   While the HH system is quantitatively satisfying, it has been relatively cumbersome and intractable for analytic investigation. Consequently, for obtaining mathematical insight into the qualitative features of nerve conduction, it is desirable to formulate simpler models. A model introduced and studied by FitzHugh [27, 29] has been quite useful in this regard. It has only two dependent variables but its structure is rich enough for solutions to qualitatively mimic many excitation-propagation phenomena. Here we will motivate the model from a particular condensation of the voltage clamp data; for a similar approach see [8]. In this condensation one exploits the fast and slow time behavior of those data and derives two current-voltage relations which we will now describe.

For each of the profiles $I_{\text{ion}}(t; v)$ in Figure 2 with $v \leqslant 100$ mV, we can identify a local minimum which occurs shortly after the clamp is applied. We denote this early transient peak by $I_p(v)$ and plot the relationship I_p versus v in Figure 5. Over most of the physiological range for v, I_p is approximately equal to the peak of the fast sodium current; it reflects the fast sodium activation m. We note two exceptions. For $v \approx \bar{v}_{\text{Na}} = 115$ mV the sodium contribution is relatively small and therefore I_p in Figure 5 for 100 mV $\leqslant v \leqslant 125$ mV is set equal to I_{ion} (of Figure 2) at the time the theoretical sodium current reaches its peak. Also, for small v, the inward sodium current is outweighed by the outward flows. The upper inset of Figure 5, which expands the region near the origin, emphasizes this latter point by showing that $I_p(v)$ is positive for small $v > 0$. Recall that positive I_{ion} does not tend to excite the membrane. Hence these positive early clamping currents for small

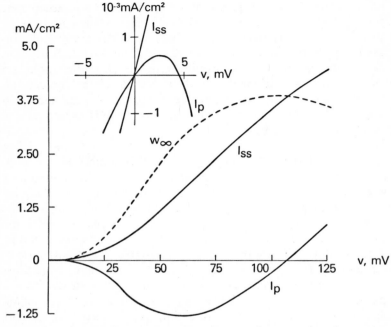

FIG. 5. Current-voltage curves from HH theoretical voltage clamp data (Fig. 2). Early transient peak current I_p and late steady state current I_{ss} are plotted versus clamp potential v. Upper inset expands region near the origin.

v are consistent with the fact that the rest state is stable for small displacements. We further remark that the N-shaped curve, I_p versus v, is qualitatively similar to the current-voltage curve which follows by setting $\tau_m = 0$, $\tau_h = \tau_n = \infty$ in (6), or equivalently $m = m_\infty(v)$, $h = h_\infty(0)$, $n = n_\infty(0)$, and evaluating the HH expression (4) for I_{ion}.

For the long time behavior, we see that for each curve in Figure 2 there is an asymptotic steady state value

$$I_{ss}(v) = \lim_{t \to \infty} I_{ion}(t; v).$$

The function $I_{ss}(v)$ is of course just I_{ion} of (4) evaluated with $m = m_\infty(v)$, $h = h_\infty(v)$, and $n = n_\infty(v)$.

Current-voltage curves such as these are frequently used by physiologists to characterize the voltage clamp data. Scott [77] illustrates curves for several different experimental preparations.

The above representation suggests that I_{ion} might be modeled as the sum of two components, one fast, denoted by f and the other slow, denoted by w. Here we suppose that f acts so rapidly that it effectively yields an instantaneous current-voltage law: $f = f(v)$. Thus we have

$$I_{ion} = f(v) + w. \tag{7}$$

For the voltage clamp configuration this means

$$I_{ion}(t; v) = f(v) + w(t; v). \tag{8}$$

We assume w satisfies an equation of the form

$$\frac{\partial w}{\partial t} = (w_\infty(v) - w)/\tau_w, \tag{9}$$

where τ_w is a constant. Because w is a slow variable, τ_w is like a typical (slow) time scale for h and n on the order (10 msec).

We now match (8) to the reduced voltage clamp data of Figure 5. This requires

$$I_p(v) = \lim_{t \to 0^+} I_{ion}(t; v) = f(v) + w_\infty(0)$$

and

$$I_{ss}(v) = \lim_{t \to \infty} I_{ion}(t; v) = f(v) + w_\infty(v).$$

With zero resting contribution for each component, $w_\infty(0) = 0$ and $f(0) = 0$, it follows from these two equations that

$$f(v) = I_p(v)$$

and

$$w_\infty(v) = I_{ss}(v) - I_p(v).$$

The N-shaped current-voltage relation, $f(v)$ versus v, mimics the coupling between m and v in the HH equation and provides the mechanism for excitability, outward current for small $v > 0$ but inward current for sufficiently large v. The "steady state" function $w_\infty(v)$ is shown dashed in Figure 5. For $v > 0$, w_∞ is positive in the physiological range of Figure 5. The slow variable w which determines the late outward component of $I_{ion}(t; v)$ mimics the effect of h and n for the HH model. Since these are responsible for returning the membrane potential to rest after impulse initiation, they, and similarly w, have been referred to as recovery variables by FitzHugh [29].

A further approximation usually made for these two variable models is to replace $w_\infty(v)$ by a term proportional to v. Finally then dimensionless variables may be introduced. Rather than carry out the scaling for the squid data we will merely proceed to the qualitative dimensionless model. Hence, with the linear replacement for w_∞ and with dimensionless variables now replacing those used above, it follows from (2), (7) and (9) that the model equation for the spatially unrestricted "axon" may be written as

$$\frac{\partial v}{\partial t} = \frac{\partial^2 v}{\partial x^2} - f(v) - w,$$

$$\frac{\partial w}{\partial t} = b(v - \gamma w). \tag{10}$$

For (10), $f(v)$ is N-shaped as shown in Figure 6, $\gamma \geqslant 0$, and $b \geqslant 0$ with usually $b \ll 1$. The constant γ should be small enough that $v = 0 = w$ is the only constant solution to (10) which thus corresponds to the unique rest state in nerve. A typical hypothesis on $f(v)$ is that in Figure 6 the horizontally hatched area exceed the vertically hatched area. This condition is analogous to the physiologists' intuitive notion of safety factor (quantitatively introduced by Rushton [74]). Roughly, it requires that adequate inward current be available to sustain impulse propagation.

In FitzHugh's presentations [27, 29], the model is derived by modifying van der Pol's relaxation oscillator. Although the equation is written differently there, (10) follows after a change of

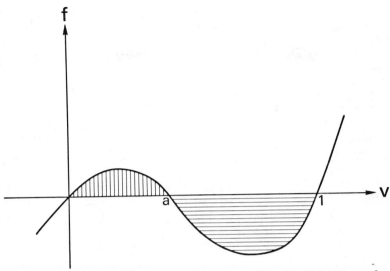

FIG. 6. Instantaneous, N-shaped, current-voltage nonlinearity, $f(v)$ versus v, for FHN equation (10).

variables; here, b plays the role of FitzHugh's temperature-like parameter ϕ. For this case, $f(v)$ is a cubic polynomial

$$f(v) = v(v - a)(v - 1) \qquad (11)$$

and the area hypothesis translates to $0 < a < \frac{1}{2}$. For such a cubic current-voltage law, Hunter, McNaughton and Noble [41] have defined a safety factor, $S = (2a)^{-1} - 1$, which must be positive for excitation and propagation. To acknowledge van der Pol and also Bonhoeffer, who earlier studied a two variable nerve analog (see [27] for references), FitzHugh referred to (10), (11) as the BVP model. He introduced the model for the space-clamped case and compared the dynamic characteristics of the cubic and HH models in addition to their I_p and I_{ss} curves.

Subsequently, Nagumo, Arimoto, and Yoshizawa [53] constructed an electronic transmission line whose behavior is described by (10). They used a tunnel diode for the nonlinear current voltage element. In addition to hardware simulations they carried out numerical calculations for the cubic case to expose the structure of the simple model. Sometimes (10) is referred to as Nagumo's equation. Here, to credit the contributions of FitzHugh and Nagumo, *et al.*, (10) will be called the FitzHugh-Nagumo (FHN) equation.

Here we have chosen particular aspects of the voltage clamp data from which to motivate the FHN equation. The choice is not unique. For example, Hunter, *et al.*, [41] have used different current-voltage curves in deriving $f(v)$ for FHN type models. They have also attempted to fit various simple functions (cubic, quintic, etc.) for $f(v)$ to data from squid nerve and cardiac Purkinje fibers. Our interest here, as was FitzHugh's, is in the qualitative features of the FHN equation as a conceptual model rather than its accuracy as a quantitative approximation.

Finally, we again state that the FHN model neglects the autonomous time behavior of Na^+ activation by effectively treating it as instantaneous. Actually, as FitzHugh points out, v and m have similar time scales. Two variable models which retain an activation variable but neglect recovery have been considered in [26, 41].

Before discussing propagation, let us consider further the space-clamped nerve and some dynamic properties more general than for the voltage clamp setup. The space-clamped FHN model is especially convenient for this since it can be analyzed in the phase plane. With an applied current source, the equations are

$$\frac{dv}{dt} = -f(v) - w + I_{app}(t),$$
$$\frac{dw}{dt} = b(v - \gamma w). \tag{12}$$

To illustrate the notion of a "voltage threshold" consider the effect of a brief current pulse for which $I_{app}(t) = Q \, \delta(t)$ where $\delta(\cdot)$ is the Dirac delta function and Q is the quantity of injected charge. This input acts only at $t = 0$ by instantaneously displacing v from rest to $v(0^+) = v_0$ where v_0 is proportional to Q. Figure 7 illustrates trajectories for various values of v_0. Each originates on the v axis since $w(0^+) = w(0) = 0$. The isoclines ($w = -f(v)$ and $w = \gamma^{-1}v$) are shown dashed; their intersection at $v = w = 0$ is the only singular point.

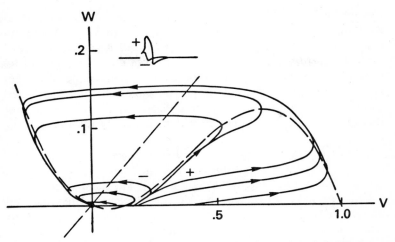

FIG. 7. Phase plane trajectories of space-clamped FHN equation (12) for an instantaneous current pulse (at $t = 0$) of various amplitudes. Each trajectory originates on the v-axis. Inset shows voltage waveforms, v versus t, for supra ($+$) and sub ($-$) threshold stimuli; the corresponding phase plane trajectories are labeled $\pm$. Model parameters are $a \approx .14$, $b \approx .008$, $\gamma \approx 2.54$. (Redrawn from FitzHugh [29].)

For small v_0, the trajectory carries (v, w) almost directly back to the origin. This is because the rest state is stable. To verify this analytically consider the linearization of (12) about the rest state:

$$\frac{dV}{dt} = -f'(0)V - W,$$

$$\frac{dW}{dt} = bV - b\gamma W.$$

It has exponential solutions like $\exp(\lambda_\pm t)$ where

$$\lambda_\pm = -(\rho + b\gamma)/2 \pm \left[(\rho + b\gamma)^2/4 - b(\rho\gamma + 1)\right]^{1/2} \quad (13)$$

and $\rho = f'(0)$; for the cubic case (11), $\rho = a$. For ρ, $b > 0$ and $\gamma \geqslant 0$, it follows from (13) that $\text{Re}\lambda_\pm < 0$.

For sizable v_0, say $.25 < v_0 < 1$, (v, w) makes a large excursion, traveling rightward and rapidly toward the cubic isocline and then upward and more slowly alongside it, before returning to rest. This feature together with stability of the rest state illustrates excitability and a threshold phenomenon. The indicated time scales become more exaggerated for smaller values of b as can be understood from a singular perturbation point of view. Trajectories for different v_0 in this range look pretty much the same after they reach the isocline. Correspondingly, physiologists refer to a space-clamped or membrane action potential even though, in general, there is not a unique trajectory which rigorously fits this description.

For a narrow range of intermediate values of v_0, $v_0 > a$ but $v_0 \approx a$, (v, w) travels up alongside the middle branch of the $v = 0$ isocline before swinging to the right or left. The maximum value of v, call it $v_{\max}(v_0)$, for each such trajectory depends quite sensitively upon v_0. For example, the inset shows $v(t)$ profiles of two trajectories, labeled $+$ and $-$, for nearly equal values of v_0.

The curve of $v_{\max}$ versus v_0 may be called the stimulus-response curve for brief current injection [29]. It is monotone increasing for $0 < v_0 < 1$ and approaches zero for $v_0 \to 0$ and one for $v_0 \to 1$. Contrary to the implication of the physiologists' "all or none" law, which precludes graded responses, the curve is not discontinuous and does not define a discrete voltage threshold. Rather it is continuous although possibly steeply sloped near $v_0 = a$. As a threshold definition in this case one could use the value of v_0 for which the slope, or sensitivity, is a maximum. FitzHugh [29] exposed this continuous threshold phenomenon for the FHN as well as the HH equations. Subsequently, experiments on squid [13] at higher temperatures to decrease sensitivity (by effectively increasing b) verified his theoretical conclusion. Of course one should realize that a sharp transition may be experimentally indistinguishable from a discontinuous jump. Consequently, the voltage at which I_{ion} (neglecting recovery) becomes inward is typically taken as the space-clamped voltage threshold (e.g., [56]); here, for the N-shaped $f(v)$, that value equals a.

Now consider a space-clamped experiment in which I_{app} is turned on at $t = 0$ and held constant, $I_{\text{app}} = I$, for $t > 0$. Under such conditions, many nerves, especially those with a sensory function, would be expected to fire repetitively for values of I in a certain interval but not for I too small or too large. Correspondingly, numerical solutions of (12) [27, 29] for appropriate values of a, b, γ indicate that for I taken small enough or sufficiently large, v and w tend to constant values. For I in an intermediate range, one observes periodic behavior which corresponds to repetitive firing. For $b \ll 1$, it is a relaxation oscillation.

Analytic results are qualitatively consistent with these features. For each constant value I for I_{app}, (12) has a unique singular point if $\gamma < 3 (1 - a + a^2)^{-1}$. In the special case $\gamma = 0$, this singular point is stable and (12) does not have a periodic solution for any I [52]. When $0 < b\gamma^2 < 1$, there are values $I_2 > I_1 > 0$ such that the singular point is linearly stable for $0 \leqslant I < I_1$ and $I > I_2$ but unstable for $I_1 < I < I_2$. The eigenvalues for the stability analysis are given by (13) with ρ replaced by $f'(v_I)$ where v_I is the steady-state equilibrium voltage. The stability transitions for $I = I_1$ or I_2 are characterized by a spiral point changing from a repellor to an attractor or vice versa [29, 81]. Hence the Hopf bifurcation theory implies that for these critical values small amplitude periodic solutions bifurcate from the steady state. Troy [81] has demonstrated that for certain parameter values these periodic solutions are stable. However a complete stimulus-response curve, oscillation amplitude versus I, has not been computed for (12) to show how these local results fit into the global picture.

Numerical results for the space-clamped HH equation also indicate repetitive firing for constant current; for example, see [12, 25, 79]. In this case, Cooley, Dodge, and Cohen [15] found that over a certain lower range of I values, the singular point is stable yet numerically one observes repetitive firing. Thus the lower critical value of I is not consistent with Hopf bifurcation of *stable* periodic solutions. However, calculations of Sabah and Spangler [75] demonstrate that the upper one is. Finally we remark that although the HH equation predicts repetitive firing, some experiments for squid nerve reveal only a finite number of pulses in response to constant current stimulation. Recently, Adelman and FitzHugh [1] have proposed a modified HH equation to account for this.

3.3. Traveling wave solutions and their stability. A traveling wave solution corresponds to a signal which propagates with constant speed and shape along a uniform nerve fiber. It is mathematically convenient since it satisfies an ordinary differential equation. As a solution to a fully specified problem for the partial differential equation it is an idealization since it satisfies only rather special boundary and/or initial conditions. It's relevance is as an asymptotic state which is attained for distances away from and/or times after stimulus application.

For the problem of determining these particular solutions there are only scattered and partial results. First, we will describe some of these results, both analytic and numerical, for the HH and FHN equations along with conjectures for the stability of the waves. Then, for a simplified, piecewise linear FHN equation which can be solved exactly, we will outline a fairly complete description of the waves and their stability.

Each of the models has a unique constant solution which is trivially a traveling wave and which corresponds to the uniform rest state in nerve. For HH, this solution is $v = 0$, $m = m_\infty(0)$, $h = h_\infty(0)$, $n = n_\infty(0)$ and for FHN, it is $v = w = 0$. Here we consider each equation as a special case of (3) with the variables identified so that $\mathbf{V} = \mathbf{0}$ is the rest solution. Consistent with the notions of excitability and threshold this constant solution is stable. Linear stability is checked in the usual way. First, linearize (3) about $\mathbf{V} = \mathbf{0}$ and then consider solutions of the form $\mathbf{U} \exp(i\alpha x - \sigma t)$ where $\mathbf{U}$ is a constant $(n + 1)$ − vector and α is real. For the HH and FHN equations, Re $\sigma < 0$ for each α and hence the rest state is linearly stable. From this follows asymptotic stability as shown by Rauch and Smoller [59] for FHN and by Evans [22] for the general class (3) if Re $\sigma \leqslant -\delta$ for some positive δ independent of α.

To mimic the conduction of a single nerve impulse, equation (3) can be expected to have a pulse-shaped traveling wave solution, $\mathbf{V}(x, t) = \mathbf{\Phi}(z)$ where $z = x - ct$ with c a constant. Here the speed c may be assumed positive since, if there is a leftward traveling wave, there is also a rightward wave. The above expectation leads to the following question. Is there a value of c such that the nonlinear autonomous ordinary differential equation

$$-c\Phi' = \begin{bmatrix} \phi_0'' \\ 0 \\ \vdots \\ 0 \end{bmatrix} + F(\Phi) \tag{14}$$

has a bounded solution which satisfies $\Phi \to 0$ as $|z| \to \infty$?

Equation (14) may be written as a first order system for the $(n + 2)$-vector $(\phi_0, \phi_0', \phi_1, \dots, \phi_n)$. Then a bounded trajectory in the $(n + 2)$-dimensional phase space for this system is equivalent to a bounded traveling wave solution to (3). In this phase space, the origin is a unique singular point and the foregoing question translates to: is there a bounded trajectory which enters the origin as $z \to \pm \infty$? Local stability analysis [22] reveals that the origin has a one-dimensional manifold Γ_+ of incoming trajectories (i.e., approaching the origin as $z \to \infty$) and an $(n + 1)$-dimensional manifold Γ_- of outgoing trajectories. The solution we seek will intersect both Γ_+ and Γ_-. For an appropriate value of c, if there is any, one can start on Γ_+ then integrate backwards (z decreasing) and eventually intersect Γ_- and thus recede into the origin. For c either too large or too small, one will not hit Γ_-, but rather head off to infinity in one or the other direction $\phi_0 \to \pm \infty$. Numerical procedures which employ this idea of "shooting" from a starting point on Γ_+ have been used to determine the HH and FHN nerve impulse solution and speed [16, 25, 29, 40, 42, 53]. With successive upper and lower estimates for c which alternate the direction of ϕ_0's approach to ∞, one gets improved approximations for c. Hodgkin and Huxley carried out this calculation by hand to achieve the results illustrated in Figure 4.

The numerical calculations also reveal that there are generally two different pulse shaped solutions. For a typical parameter (e.g., temperature) of the model, there is a double-branched speed curve. Figure 8 (right) illustrates a pulse speed curve for a simplified FHN equation; speed is plotted versus a "threshold-like" parameter, a. The nonuniqueness for $a < a_\nu$ contrasts with the uniqueness of the nerve impulse, the pulse seen on Nagumo's line, and the pulse observed in numerical solutions of the HH and FHN equations for appropriate initial-boundary value problems. It is conjectured [16, 29, 42, 53] that the slow pulse is unstable and not experimentally observable. While analytic proofs are yet to appear, recent numerical stability calculations [24a] support the conjecture for HH.

To correspond to steady repetitive firing there should also be periodic wave train solutions of (3): $V(x, t) = \Phi(z)$ with $z = kx - \omega t$ where Φ is 2π-periodic. Here k and ω are positive constants; k is the wave-number, proportional to the reciprocal of wavelength or pulse spacing, and ω is proportional to firing frequency. The propagation speed c equals ω/k. Such a periodic solution to (3) satisfies

$$- \omega\Phi' = \begin{bmatrix} k^2\phi_0'' \\ 0 \\ \vdots \\ 0 \end{bmatrix} + F(\Phi) \tag{15}$$

and $\Phi(z) = \Phi(z + 2\pi)$ and is of course equivalent to a periodic solution to (14) with period $2\pi/k$ and speed ω/k.

For the different frequencies of nerve firing there should be a one parameter family of periodic wave trains. Numerical solutions of initial-boundary value problems corresponding to maintained stimulation by an electrode reveal such periodic wave propagation [16, 79]. The full range of wave train solutions and appropriate values of k and ω for the HH and cubic FHN equations have not yet been determined by solving (15). This has been done for the simplified FHN equation [69]. I will describe the results and discuss the stability of the wave trains.

Yet another class of traveling wave solutions is motivated by the following consideration. Suppose a nerve is stimulated successively with a few brief suprathreshold current injections and suppose it fires N pulses sequentially. As these N pulses propagate will they eventually position themselves to achieve some equilibrium spacing and then all steadily propagate at one fixed velocity? Correspondingly, is there a value $c = c_N$ such that (14) has an N-pulse solution, i.e., a bounded solution with $\Phi \to 0$ as $|z| \to \infty$ and for which ϕ_0 looks like a train of N action potentials? To carry this a step further, suppose such a stimulus burst is presented periodically with a rest interval between burst presentations. One could then ask whether there are corresponding burst type periodic wave trains with N action potentials per wavelength. These possibilities would suggest an even richer analytic structure than described above. However these latter questions are perhaps more of mathematical than physiological interest. Certainly it would seem so for nerves, such as squid axon, which are not long relative to pulse width and for which detailed inter-pulse timing would have no apparent functional significance. Furthermore even for long nerves one might argue, considering biological noise, that precise pulse spacing would not be maintained during propagation.

Analytic results for traveling wave solutions are based upon hypotheses which emphasize the different time scales in the equations. As an extreme special case, the recovery variables are taken to be infinitely slow ($\tau_h = \infty = \tau_n$ for HH; $b = 0$ for FHN) and are held constant at their resting values. Each reduced model has a traveling wave solution in the form of a front rather than a propagating pulse [9, 29, 52, 53]; Hunter, *et al.*, [41] have determined the front speed analytically for several explicit forms of $f(v)$ in the FHN equation. The front represents a propagating transition from rest to an excited state with no recovery back to rest. When the recovery variables are included as very slow, the speed of the fast (stable) pulse is close to that of the front solution to the reduced model. For the cubic FHN equation with $b > 0$, $\gamma = 0$, Casten, Cohen, and Lagerstrom [8] exploited $b \ll 1$ and used the singular perturbation technique of matched asymptotic expansions to obtain the fast and slow solitary pulses. They further obtained periodic wave trains of large wavelength and small frequency with speed close to that of the fast pulse. Existence proofs for solitary pulse and periodic traveling wave solutions to the FHN and restricted forms of the HH equations have been given by Carpenter [7, 7a], Conley [14], and Hastings [35, 36] (see [37] for an introductory presentation of Hastings' analysis). For the FHN equation, such results verify the necessity of the area hypothesis for $f(v)$. In [7], Carpenter also demonstrates that an HH type equation may admit N-pulse traveling wave solutions. Her investigation suggests that the FHN equation, because it has only one recovery variable, does not admit such N-pulse solutions for $b \ll 1$. While some existence proofs apply only to the fast stable pulse, analytic results of Evans [24] indicate that if a nerve conduction equation has a stable solitary pulse solution then it also has an unstable one.

A solvable model equation. The above structural features, a stable rest state, solitary pulse and periodic traveling waves, qualitatively characterize a class of nerve conduction models. Yet a full parametric description of the traveling wave solutions to either the FHN or HH equations has not been provided. With this in mind, McKean [52] suggested an FHN caricature to further ease the analytic burden. In (10), he set $\gamma = 0$ and chose for $f(v)$ the piecewise linear function

$$f(v) = v - H(v - a), \quad 0 < a < \tfrac{1}{2}, \tag{16}$$

where $H(\cdot)$ is the Heaviside step function. For this choice, (10) has piecewise constant coefficients. Different piecewise linear models for nerve conduction were earlier studied by Rushton [74] and others (e.g., see [29]). As a conceptual analogue for excitable chemical reaction-diffusion systems, Winfree [82] has used a variation of this simplified FHN model. Here we will describe the traveling wave solutions to (10), (16); they were obtained explicitly by Rinzel and Keller [69]. Then we will discuss their stability: both temporal stability, following Rinzel and Keller, and spatial stability, as considered by Rinzel [70].

First we discuss the solitary pulse case. A traveling wave solution $v(x, t) = v_c(z)$, $w(x, t) = w_c(z)$, $z = x - ct$ satisfies

$$\begin{bmatrix} v_c \\ v_c' \\ w_c \end{bmatrix}' = \begin{bmatrix} 0 & 1 & 0 \\ 1 & -c & 1 \\ -b/c & 0 & 0 \end{bmatrix} \begin{bmatrix} v_c \\ v_c' \\ w_c \end{bmatrix} - \begin{bmatrix} 0 \\ H(v_c - a) \\ 0 \end{bmatrix}. \tag{17}$$

The profile of a rightward moving v_c-pulse is shown in Figure 8 (left). Because (17) does not depend explicitly on z the origin $z = 0$ may be chosen so that $v_c(0) = a$. Solutions to (17) are composed of sums of exponentials $\mathbf{X}_i \exp(\alpha_i z)$, $i = 1, 2, 3$, where the α_i and $\mathbf{X}_i$ are the eigenvalues and eigenvectors of the 3×3 matrix in (17). The α_i satisfy

$$\alpha_1 < 0; \quad \mathrm{Re}\, \alpha_2, \quad \mathrm{Re}\, \alpha_3 > 0.$$

Note that in the context of our above discussion, $\mathbf{X}_1$ spans the manifold Γ_+ and $\mathbf{X}_2$, $\mathbf{X}_3$ span Γ_- for $v_c < a$ in the phase space of (17). Hence, the unique trajectory which satisfies $v_c(0) = a$ and which enters the origin $(0, 0, 0)$ as $z \to \infty$ is given by $a\mathbf{X}_1 \exp(\alpha_1 z)$. Here, the first component of $\mathbf{X}_1$ is normalized to one. By matching exponentials at $z = 0$, this solution can be extended through the interval $z_1 < z < 0$. Again by matching at z_1, the solution is defined for $z < z_1$. To satisfy the condition that the solution tend to zero as $z \to -\infty$ we set to zero the coefficient of $\mathbf{X}_1 \exp(\alpha_1 z)$. This leads to a scalar transcendental equation for the speed c in terms of a and b. For fixed b, the solution c, obtained numerically, in terms of a yields a double branched speed curve; a typical case is shown in Figure 8 (right).

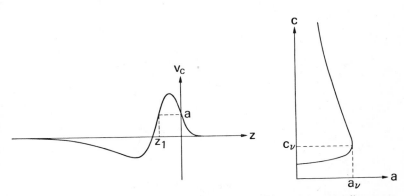

FIG. 8. "Membrane potential" profile (left) for solitary pulse traveling wave solution $v_c(z)$, $w_c(z)$ of equations (10), (16) where $z = x - ct$. The traveling pulse also satisfies equation (17). Pulse speed curve (right) illustrates propagation speed c versus a for $b = .05$, $\gamma = 0$; here, $a_\nu = .35$, $c_\nu = .43$.

For a periodic wave train we seek a solution to (10), (16) of the form $v(x, t) = v_c(z)$, $w(x, t) = w_c(z)$, $z = kx - \omega t$, where v_c, w_c are 2π-periodic. The desired v_c profile is shown in Figure 9 (left). Such a solution is constructed by patching exponentials in a manner similar to that above. In this case, the matching conditions and periodic boundary conditions lead to a transcendental equation which involves the wave-number k, the frequency ω, and the model parameters a and b. For fixed a and b, this equation was solved numerically to produce a dispersion relation $\omega = \omega(k, a, b)$. A particular case is illustrated in Figure 9 (right). For each frequency ω less than some maximum $\omega_{\max}$ there are two periodic solutions with different wave-numbers and hence different speeds. The right branch of the dispersion curve corresponds to the slower wave trains. The slope $d\omega/dk$ of the fast (slow) branch tends, as $k \to 0$, to the speed of the fast (slow) solitary pulse. Correspondingly, a periodic wave train tends to a solitary pulse as the wavelength tends to infinity. This limiting behavior was also exposed for the cubic FHN model by Casten, *et al.*, [8].

The above results suggest that in general there is a dispersion curve for the periodic wave train solutions to a nerve conduction equation. An interesting physiological consequence of dispersion is

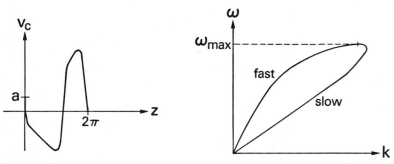

FIG. 9. "Membrane potential" profile (left) for periodic traveling wave solution $v_c(z)$, $w_c(z)$ of equations (10), (16) with $z = kx - \omega t$ and period 2π. Dispersion curve (right) shows frequency ω versus wave-number k for $a = .3$, $b = .05$, $\gamma = 0$; here $\omega_{\max} = .129$.

that during repetitive activity the propagation speed depends upon the steady firing rate. The relative variation in speed will depend upon the particular axon. More generally, for a non-periodic stimulus at $x = 0$, the temporal sequence of impulse firings will define a time-varying "instantaneous" frequency. Then during propagation, if dispersive effects are significant, the relative temporal pattern will be altered and the instantaneous frequency at a distant location will not be merely a delay of that at the source end. In this case, irrespective of biological noise or axon structural variations, one would anticipate that some temporal average of frequency is of greater functional significance than individual pulse arrival times. Dispersive aspects of nerve conduction have also been discussed by [3, 31].

Now we turn to the stability of the waves. One way to analyze the stability of an equilibrium solution is to study the temporal evolution of small perturbations. This approach is appropriate for initial value problems. Since the equilibrium solutions of our concern are traveling waves, we introduce a traveling coordinate frame $t, z = kx - \omega t$. Here $t \geqslant 0$ and $-\infty < z < \infty$. The wave is a t-independent solution in this coordinate system. For linear stability, we consider the linear variational equation

$$\frac{\partial \tilde{v}}{\partial t} = k^2 \frac{\partial^2 \tilde{v}}{\partial z^2} + \omega \frac{\partial \tilde{v}}{\partial z} - f'(v_c)\tilde{v} - \tilde{w},$$

$$\frac{\partial \tilde{w}}{\partial t} = \omega \frac{\partial \tilde{w}}{\partial z} + b\tilde{v}. \tag{18}$$

For a solitary pulse solution we formally set $k = 1$ and $\omega = c$ in (18).

Now we seek solutions to (18) of the form

$$\tilde{v}(z, t) = e^{\mu t} v_T(z), \quad \tilde{w}(z, t) = e^{\mu t} w_T(z),$$

and we are led to an eigenvalue problem for v_T, w_T, and μ:

$$-A\mathbf{V}_T = \mu\mathbf{V}_T, \tag{19}$$

where

$$
- A \equiv \begin{bmatrix} \left(k \dfrac{d}{dz} \right)^2 + \omega \dfrac{d}{dz} - f'(v_c) & -1 \\[2ex] b & \omega \dfrac{d}{dz} \end{bmatrix} \tag{20}
$$

and

$$
\mathbf{V}_T = \begin{pmatrix} v_T \\ w_T \end{pmatrix}.
$$

If, for some μ with $\mathrm{Re}\,\mu > 0$, (19) has a bounded solution $\mathbf{V}_T$ in an admissible class of functions then (v_c, w_c) is temporally unstable; $\mathbf{V}_T$ is an unstable mode with growth rate $\mathrm{Re}\,\mu > 0$. Because (19) has piecewise constant coefficients, solutions can be constructed as sums of exponentials. Note for our discontinuous $f(v)$ we have $f'(v_c) = 1 - \delta(v_c - a)$ and in (19) this means that dv_T/dz satisfies a jump condition wherever $v_c = a$. By requiring that a particular solution of (19) satisfy certain boundary conditions, we are led to a transcendental equation which involves μ, the wave parameters k and ω, and the model parameters a and b. This "characteristic" equation may be solved numerically.

In this way it was demonstrated [69] that the slow solitary pulse solution is temporally unstable. A positive growth rate $\mu > 0$ was found corresponding to an unstable mode $\mathbf{V}_T$ which tends to zero as $|z| \to \infty$. Hence the slow branch of the pulse speed curve represents unstable solutions; Figure 10 (left) illustrates this. The unique pulse for a_ν, c_ν is neutrally stable. The distinguishing eigenvalue μ passes through zero as the speed curve knee is rounded and μ becomes negative on the upper branch.

Since the slow pulse is unstable, it is natural to conjecture that wave trains corresponding to some portion of the slow branch of the dispersion curve are likewise unstable. Rinzel and Keller also analyzed the temporal stability of these solutions. They calculated a positive growth rate $\mu > 0$ for a 2π-periodic unstable mode $\mathbf{V}_T$ corresponding to each slow wave train on some portion of the dispersion curve. For each set of values a and b which they

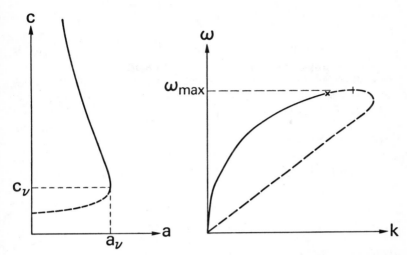

FIG. 10. Solitary pulse speed curve (left) from Fig. 8; dashed portion indicates temporally and spatially unstable pulse solutions. Neutral stability for a_ν, c_ν. Dispersion curve (right) for periodic wave trains with $a = .25$, $b = .005$, $\gamma = 0$; dashed portion indicates temporally unstable wave trains. For spatial stability, right branch only would be dashed; neutral spatial stability for ω_{max}.

considered, the long wavelength slow waves are unstable. In some cases, the connected segment of the dispersion curve which corresponds to instability may include the entire slow branch and some upper portion of the fast branch. This means that the transition from stability to instability does not necessarily occur for $\omega = \omega_{max}$. Figure 10 (right) illustrates a particular example of this.

An alternative notion of stability is suggested by the following consideration. The configuration of a nerve fiber stimulated at a fixed location is typical for a neuroelectric signaling problem. The stimulating current might be applied externally with an electrode or it might be supplied through the nerve cell body as a result of dendritic synaptic activity. For a periodic stimulus, a simple mathematical formulation might neglect the transient behavior, just after the stimulus is turned on, and consider only the steady repetitive firing aspect. This means solving a boundary value problem with data imposed at a fixed location, $x = 0$ say, and specified for all time, $-\infty < t < \infty$. For appropriate data, one expects to observe, at large values of x, a stable periodic wave train.

The relevant notion of stability in this signaling context is spatial rather than temporal. For this, we are called upon to examine the growth with distance along the nerve of perturbations imposed on the waveform at $x = 0$ and specified for all time. We remark that our intuitive motivation for spatial stability would also be relevant in other biological and physical wave signaling problems. The appropriate coordinate frame for our analysis is x, $z = kx - \omega t$ where $x \geqslant 0$ and $-\infty < z < \infty$. In these coordinates, the traveling wave is an x-independent solution. The linear variational equation is

$$
\begin{aligned}
- \omega \frac{\partial \tilde{v}}{\partial z} &= k^2 \frac{\partial^2 \tilde{v}}{\partial z^2} + 2k \frac{\partial^2 \tilde{v}}{\partial z \partial x} + \frac{\partial^2 \tilde{v}}{\partial x^2} - f'(v_c)\tilde{v} - \tilde{w}, \\
- \omega \frac{\partial \tilde{w}}{\partial z} &= b\tilde{v}.
\end{aligned}
\tag{21}
$$

Here we seek solutions of the form

$$
\tilde{v}(z, x) = e^{\lambda x} v_S(z), \quad \tilde{w}(z, x) = e^{\lambda x} w_S(z).
$$

This leads to an eigenvalue problem for v_S, w_S, and λ:

$$
A\mathbf{V}_S = \lambda D(\lambda)\mathbf{V}_S
\tag{22}
$$

where A is given by (20), and $D(\lambda)$ is given by

$$
D(\lambda) = \begin{bmatrix} \lambda + 2k \dfrac{d}{dz} & 0 \\ 0 & 0 \end{bmatrix},
\tag{23}
$$

and

$$
\mathbf{V}_S = \begin{pmatrix} v_S \\ w_S \end{pmatrix}.
$$

The wave train is spatially unstable if, for some λ with Re $\lambda > 0$, equation (22) has a bounded solution $\mathbf{V}_S$ belonging to an admissible class of perturbations. The solution $\mathbf{V}_S$ is an unstable mode with growth parameter Re λ. Just as for temporal stability, the eigenvalue problem involves an equation (22) with piecewise constant coefficients. As in that previous case, by demanding that a particular solution satisfy certain boundary conditions we are led to a transcendental equation for λ which involves k, ω, a, and b.

By solving this equation numerically, Rinzel [70] has demonstrated that the slow pulse and the slow wave trains are spatially unstable solutions to the simplified FHN equation. The solitary pulse for a_ν, c_ν has neutral spatial stability in agreement with the temporal stability results; Figure 10 (left) applies in this case. For the periodic wave trains, the wave with $\omega = \omega_{max}$ has neutral spatial stability for each set of values a, b which were considered. This shows, in contrast to temporal stability, that neutral spatial stability for the wave trains has a crisp characterization in terms of the dispersion curve. A diagram analogous to Figure 10 (right) to illustrate spatial stability would have only the right branch dashed.

Certain stability results for the simplified FHN equation have suggested their extension to a general class of equations of the form (3). For the solitary pulse solutions, Rinzel [71] has formally shown that neutral stability, temporal and spatial, corresponds to the speed curve knee. In addition, neutral spatial stability for the periodic wave trains should occur at local maxima or minima on the dispersion curve. These results, as well as those for the simple model, are for linear stability. It is assumed that linear stability determines asymptotic stability; rigorous verifications of this for temporal stability have been given by Evans [21, 22, 23, 24] for a solitary pulse solution to (3), and by Sattinger [76] for the traveling front solution to FHN without recovery.

3.4. Stimulus-response properties.

In the preceding section we considered steady propagation and put aside the matter of initiation of a solitary pulse or a train of pulses in response to particular stimuli. Here our goal is to formulate problems and characterize data for which the traveling waves are asymptotic solutions. A complete characterization is equivalent to specifying the domains of attraction for each of these special equilibrium states. Our preceding linear stability results indicate what to expect locally around these states. However global attractive characteristics are

more difficult to describe and to date few rigorous results exist.
The results we will describe are primarily numerical and are for
particular, but experimentally relevant, stimulus configurations.

Consider a uniform axon of infinite length which is stimulated
from rest by an electrode supplying current at $x = 0$. For this we
have (3), using either HH or FHN dynamics, with $-\infty < x < \infty$,
$t \geqslant 0$, $\mathbf{V}(x, 0) = 0$ and a source term proportional to $I_{\text{app}}(x, t) =
I_{\text{app}}(t)\delta(x)$ in the first of (3). By symmetry, this problem is
equivalent to specifying $\partial v/\partial x$ (proportional to $I_{\text{app}}(t)$) at $x = 0$
for a semi-infinite axon, $x \geqslant 0$. Here our discussion will be for this
latter formulation; it is also relevant for current supplied from the
cell body. We will assume a typical form for $I_{\text{app}}(t)$, a constant
current source I turned on at $t = 0$ and turned off at $t = t_1$:

$$I_{\text{app}}(t) = IH(t)H(t_1 - t). \tag{24}$$

First we consider the excitation of a solitary pulse and ask: For
a given stimulus strength I, what is the minimum duration $t_1 = t_*$
sufficient for the initiation of a single impulse? To rephrase this,
let Φ_h denote the translated stable pulse $\Phi_h(x - ct) = \Phi(x -
ct + h)$. Then, to answer the question, determine $t_* = \inf_{t_1 > 0} t_1$ (if it
exists) such that $\lim_{t \to \infty} \|\mathbf{V}(\cdot, t) - \Phi_h(\cdot)\| = 0$ for some constant h.
The norm here, for an $(n + 1)$-vector $\mathbf{Y}(x, t)$, could be taken (for
example) as $\max_{0 \leqslant i \leqslant n} \sup_{x \geqslant 0} |y_i(x, t)|$.

By numerical solution of the above initial-boundary value problem, the theoretical strength-duration (S-D) curve, I vs. t_*, has
been computed for the HH equation [16, 56]. The data points $(+)$
in Figure 11 were obtained by such calculations. (The data points
(x) are for the corresponding space clamped case.) Note that this
computational approach, as the experimental setup, confines one
to finite axon length and integration time. However, investigators
report no difficulty identifying steady propagation. For reasonable
lengths, well shaped action potentials are usually observed within
a distance of one pulse width from the source. For the computations, the boundary condition imposed at the far end is usually
$\partial v/\partial x = 0$ (sealed end) or $v = 0$ (clamped end). The numerical
results agree qualitatively with experimental curves and the following features.

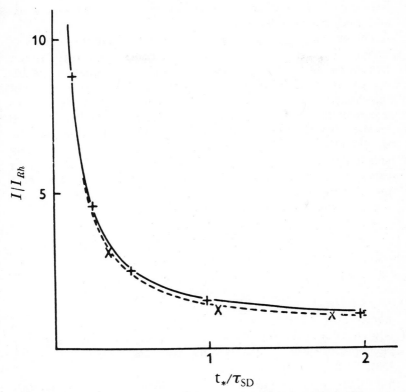

FIG. 11. Strength-duration relation, I versus t_*, for the HH equation with current stimulation at a single point $x = 0$ (data points, $+$) and uniformly over a space-clamped region (data points, x). The curves are limiting cases for Hill's two variable space-clamped theory of excitation. (Redrawn from Noble and Stein [56].)

For an intense stimulus only a short duration is required. In this case, the condition for excitation is evidently that the applied charge $t_1 I$ be not less than a critical value Q_{Th} where, according to the S-D curve, $Q_{Th} = \lim_{t_* \to 0} t_* I$. On the other hand, a very weak current, $I < I_{Rh}$, will not generate a pulse even if $t_1 = \infty$; I_{Rh} is called the rheobase value. These empirical threshold values are used to scale $I(I/I_{Rh})$ and $t_*(t_*/\tau_{SD}$ where $\tau_{SD} = Q_{Th}/I_{Rh})$ as in

Figure 11. Hence dimensionless S-D curves for different axons and models may be compared.

The curves in Figure 11 are according to a simplified theory of excitation proposed variously by Hill [38] and others. It is a simple linear description of subthreshold space-clamped behavior and has two dynamic variables: potential and "threshold." By allowing the threshold to depend upon time, the model mimics recovery or accommodative effects. "Excitation" occurs when the potential reaches threshold; however a dynamic description beyond threshold is not prescribed. Figure 11 illustrates two limiting cases which bound all the S-D curves derived from the theory. For example, the upper curve is obtained for the special case of a model with fixed threshold, a. With $v(0) = 0$ and t_* equal to the time for which $v(t) = a$, it follows from

$$v_t = -v + IH(t)$$

that

$$I/I_{Rh} = \left[1 - \exp(-t_*/\tau_{SD}) \right]^{-1}$$

where $\tau_{SD} = 1$ and $I_{Rh} = a$. For additional discussion and related references see [29, 56]. While Hill's theory might be excessively phenomenological, it qualitatively describes S-D data even for spatially non-uniform stimulation.

For the more relevant HH or FHN models, the S-D curve or even its asymptotic behavior awaits rigorous analytical derivation. However, for reduced versions of these equations certain threshold properties have been considered. The reductions typically exaggerate fast and slow time scales and one considers an FHN equation without recovery

$$\frac{\partial v}{\partial t} = \frac{\partial^2 v}{\partial x^2} - f(v) \tag{25}$$

for $x > 0$, $t > 0$. Recall that (25) has a stable traveling wave solution which is a front rather than a pulse. In this case, excitation means $\lim_{x \to \infty} \lim_{t \to \infty} v(x, t) = 1$.

Recently, Aronson and Weinberger [2] have rigorously demonstrated a threshold effect for (25). In their treatment the potential $v(0, t)$ at $x = 0$ is given rather than the current. To specify $v(0, t)$ corresponds to dynamic voltage clamping at a single point; the feasibility of its implementation (not considered in [2]) is a partially open matter and leads to some interesting stability questions. Now to state some results of Aronson and Weinberger we define, according to the area hypothesis, a value κ such that $a < \kappa < 1$ and $\int_0^\kappa f(v)\, dv = 0$. Under the assumptions $v(x, 0) = 0$ and $0 \le v(0, t) \le 1$, they show: (i) if $v(0, t) \le \kappa$, then excitation cannot occur; while (ii) if $v(0, t) > \kappa$ for sufficient duration, then excitation must occur. These results and others are based upon the maximum principle and methods which exploit certain t-independent solutions to (25) as comparison functions. A particularly useful one $\psi(x)$ is monotone decreasing and satisfies $\psi(0) = \kappa$, $\lim_{x \to \infty} \psi(x) = 0$.

For a biophysical interpretation, recall that the "membrane" generates inward, or depolarizing, ionic current whenever or wherever $f(v) < 0$. Thus, since $f(v) < 0$ for $a < v \le \kappa$, result (i) shows that even though the "axon" may generate inward current near $x = 0$, and for an arbitrary period of time, excitation is not guaranteed. This contrasts with the corresponding space-clamped version of (25) for which excitation ($v(t) \to 1$) occurs if v ever rises above the level a.

Physiologists had recognized early the notion that excitation by a point source requires a finite and adequate quantity of inward ionic current, i.e., just raising $v(0, t)$ to the "space-clamped voltage threshold" is insufficient. Attempts at quantifying and applying threshold concepts have been based on formal, intuitive arguments. Rushton [74], in a classical paper, suggested that when a current input is turned off a minimum length of axon must be generating inward current to insure excitation. For a simple model, he calculated a so-called threshold or liminal length x_{LL}. Fozzard and Schoenberg [32] applied this concept to describe S-D data for cardiac Purkinje fibers. By assuming values of x_{LL} and considering excitation to occur when $v(x_{LL}, t) = a$, they solved (25) (linearized about $v = 0$) to approximate the necessary duration. To obtain an estimate for x_{LL}, Noble [57] employed $\psi(x)$, the steady state solution defined above. He proposed the value of x

for which $\psi(x) = a$ and referred to κ as the voltage threshold for excitation by a point source. Other interpretations have been given to $\psi(x)$ in the S-D context: Noble and Stein [56] obtained an upper estimate for Q_{Th} and Jack, *et al.* [47], approximated I_{Rh}. We can hope that continued theoretical effort will bring into sharper focus the mathematical mechanism for excitation. Perhaps the appropriate functional of $I_{app}(t)$ or $v(0, t)$ and its threshold values will be determined and interpreted to provide additional biophysical insight.

We now consider the signaling problem for periodic trains of pulses in which the stimulating current is left on indefinitely; $t_1 = \infty$ in (24). Numerical solutions [16, 79] to the initial-boundary value problem for the HH equation (6) suggest some analogies with the space clamped case of repetitive firing. There is a finite interval of values for I such that steady firing is observed. Here steady firing means that, after an initial transient period, the solution is time periodic and, for reasonable distances from the source, $v(x, t)$ looks like a periodic train of steadily propagating action potentials. Note that the theoretical measure for convergence to a periodic wave train would be different from that for a pulse as introduced earlier in this section. Figure 12 illustrates the calculated voltage time course at two positions, $x = 0$ and $x = 1$ cm., with four different values of I.

The corresponding firing frequency (pulses/sec.) versus current strength I is shown in Figure 13 for (6) and two modified versions of (6). The modifications correspond to translation of the curves in Figure 3 by 5 or -10 mV and they are to model changes in calcium concentration. At a lower critical value of I, the frequency jumps from zero to a finite value. Then typically it increases with I over a certain range, i.e., a stronger stimulus generally means a faster firing rate. At higher critical values of I the frequency evidently drops discontinuously, e.g., by 1/2 (Stein, private communication). For I too large, there is no repetitive activity; only a finite train of pulses is generated during the initial transient and then the solution apparently tends to a t-independent steady state. These current-frequency curves for point stimulation are qualitatively similar to experimental curves for a variety of neurons although such data for squid axon is apparently not available [47].

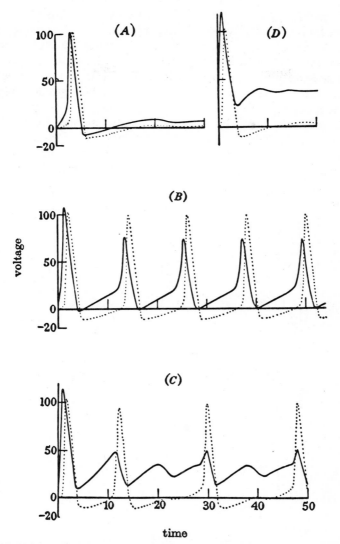

FIG. 12. Voltage response, $v(x, t)$ versus t, at $x = 0$ (solid) and $x = 1$ cm. (dotted) for HH equation (6) with step of current applied at $x = 0$. The current strengths are: (A) 0.76 μA, (B) 2.53 μA, (C) 6.1 μA, (D) 10 μA. Cases (B) and (C) illustrate repetitive firing. (From Stein [79] with permission of the Royal Society.)

At their low I ends, the curves do not show firing at very low frequencies. Evidently, low frequency wave trains are not generated in this way for this stimulus configuration and time course.

As in the space-clamped case, some analytic results are consistent with this qualitative picture. To illustrate, we neglect the transient aspects of the signaling problem and consider only the time independent or steady periodic behavior. For each of the equations we have considered there is a spatially non-uniform t-independent solution for point stimulation by a constant current, $I_{app}(x, t) = I \delta(x)$, for any value of I. We remark that the FHN case is interesting only if $\gamma > 0$; for $\gamma = 0$, repetitive firing appar-

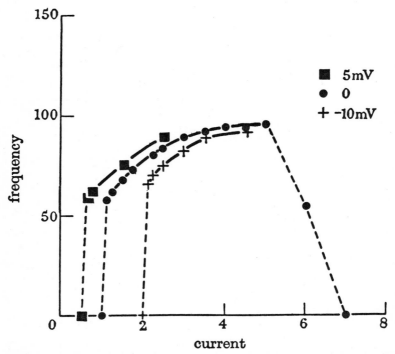

FIG. 13. Steady firing frequency (pulses/sec) versus current strength (μA) for HH equation (6) and two modifications (see text) with current stimulus at $x = 0$. (Redrawn from Stein [79] with permission of the Royal Society.)

ently does not occur. Our intuition suggests that this steady state is unstable for a finite interval of I values; this interval should be a subinterval of the one for repetitive activity. Explicit analysis (Rinzel [73a]) for a simplified "sawtooth" FHN equation with appropriate parameter values indeed reveals that this steady state is temporally unstable for an interval of I values. Bifurcation of time periodic solutions is found at the upper and lower critical values of I. Moreover, it can be seen formally for the general FHN equation that stability transitions of the steady state are associated with bifurcation of time periodic solutions. This analytic structure is consistent with the curves of Fig. 13 and the jump discontinuities in frequency.

Remarks on axonal structural variations.   Up to this point we have assumed a theoretical axon with uniform characteristics. Structural variations in either membrane or geometrical properties were not considered. Here for particular cases we will briefly discuss some consequences for signaling phenomenology.

First we consider the case of a myelinated or noded axon. Such a fiber is sheathed by an insulating layer of myelin. Excitable membrane is exposed only at regularly spaced nodes or gaps in the myelin. Although such axons expose only a small fraction of membrane area to the ionic environment, they generally conduct impulses. The action potential appears to jump from node to node. We remark that myelination is not atypical; most motor and sensory axons of vertebrates are noded.

The mathematical models typically assume spatial uniformity at each node. For the jth node excitable membrane dynamics are described by a vector $\mathbf{V}^j$ whose components satisfy a set of ordinary differential equations (e.g., HH type space-clamped equations). For the comparatively long internode segments constant membrane and sheath conductances are assumed. The internode potential distribution $v(x, t)$ satisfies a linear equation, (2) with I_{ion} proportional to v and $I_{\text{syn}} = I_{\text{app}} = 0$. Continuity of potential and conservation of current provide matching conditions at the nodes. The resulting set of coupled differential equations is semi-discrete in space. For steady propagation, one would seek τ_d, the node-to-node time delay, such that there is a bounded solution of the form $v(x, t) = v(x - l, t - \tau_d)$ and $\mathbf{V}^j(t) = \mathbf{V}^{j-1}(t - \tau_d)$ where l is the internode spacing. Propagation speed equals l/τ_d. By numerical solution of initial boundary value problems for the noded model, impulse propagation speed and its dependence upon parameters has been calculated [28, 33]. Repetitive firing has apparently not been treated. Along with other interesting features, one finds that pulse velocity is maximized for an optimal value of l or an optimal value of myelin thickness. This leads to an obvious evolutionary interpretation. For further discussion and references on myelinated axonal signaling see [28, 33, 47, 77].

Such periodic patterns of variation may be contrasted with less symmetrical ones such as a sudden change in axon diameter or an axonal branching. Calculations for explicit models (e.g., see [34, 49, 58] and their references) illustrate features which might reasonably be taken as qualitative. For example, suppose the diameter for $x < x_0$ is d_- and for $x > x_0$ it is d_+, and suppose an impulse is approaching from the left ($x = -\infty$). For $d_+ < d_-$, the impulse will successfully propagate past x_0 to $x = +\infty$. For $d_+ \gg d_-$, the impulse propagates only a short distance into $x > x_0$ before it fails. Intuitively, the substantially increased area of membrane per unit length (for $x > x_0$) requires more current for excitation than the approaching pulse supplies. Consequently, the impulse dies. For an intermediate increase in diameter, a rather interesting phenomenon can occur. As the impulse approaches x_0 from the left, it slows. As it lingers near x_0, the membrane for $x < x_0$ will be recovering from its refractoriness and regaining its excitability. If the delay at x_0 is sufficient, then the depolarization around x_0 may re-excite $x < x_0$. In this case, one would observe a pulse propagating back toward $x = -\infty$ in addition to one propagating toward $x = \infty$. Figure 14 illustrates such an example from the calculations of Goldstein and Rall [34]. Their simulations were for a three-variable nerve conduction equation (for (3), $n = 2$) which has polynomial nonlinearities. The above qualitative features also apply for an impulse which approaches an axon branching if all the daughters have identical membrane properties.

The "reflection" phenomena carry implications for the case of repetitive firing. First, note that for a uniform axon, impulses

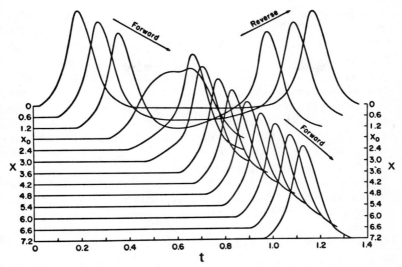

FIG. 14. Forward and reverse ("reflected") propagation of an action potential in an axon with a diameter step increase at x_0. The voltage time course, $v(x, t)$ versus t, is shown for successive locations. Solutions were obtained numerically for a three variable model. For this figure, $d_+/d_- = 2.5$. (From Goldstein and Rall [34].)

annihilate each other during head-on collisions. Now suppose for the above non-uniform case that an approaching pulse gives rise to a "transmitted" and also a "reflected" pulse. The latter would then collide with and annihilate a subsequent approaching pulse. Hence one would observe frequency demultiplication; one pulse transmitted to $x = \infty$ for every two which approach from $x = -\infty$.

4. NEURONAL DENDRITIC INTEGRATION

4.1. A mathematical model. Integration here refers to the spatio-temporal development of membrane potential throughout the cell body and dendritic trees in response to the inputs they receive. Synaptic input is generally assumed to result from membrane permeability changes at the synapse. Physiologists have identified two types of "chemical" synapse. An excitatory synaptic input produces a brief positive deviation of membrane potential from rest, an EPSP (excitatory post synaptic potential). An inhibitory one results in a negative deviation, an IPSP. For a given permeability change, the amplitude of an EPSP at a single active synapse (a "unitary" event) will depend upon its location. For a motoneuron somatic synapse, the "unitary" EPSP is on the order of $0.1 - 1.0$ mV [43, 44] with a typical "time to peak" of $0.1 - 1.0$ msec. With sufficient net excitatory input, an impulse is initiated (usually near where the axon joins the soma) and then this impulse propagates along the axon. We will not include here a mathematical description for impulse generation.

As outlined in section 2, we will assume a linear description for the transmission of potentials in the dendrites. The membrane conductance is assumed independent of potential. Synaptic potentials are said to spread passively in the dendrites. This means that the EPSP amplitude generated at a synapse is transmitted to other dendritic locations in an attenuated manner. The peak amplitude, i.e., maximum over time at a given location, decreases with distance from the input site.

We note that these assumptions are not based upon direct observation of dendritic potentials, but rather upon recordings taken at the cell body. For most cells, physiologists cannot provide intracellular electrical data from dendrites; the branches are too small to penetrate consistently.

One task for a model, which is based on such assumptions, is to account for these data from the soma. As applied to motoneuron data, Rall's model has reasonably made this accounting for a number of investigators [4, 5, 43, 44, 51, 62]. Moreover, the model theoretically demonstrates the functional significance of dendritic synaptic activity, including that which originates at distal (outlying) dendritic locations from the cell body. It is concluded that dendritic cable properties play a major role in determining neuronal integrative behavior [5, 61]. This is consistent with the anatomical observation that most synaptic contacts are distributed over the extensive dendritic surface. While such conclusions are now widely accepted, they offered a challenge to an earlier hypothesis that distal dendritic inputs were functionally insignificant [19, 20]. This earlier point of view was based upon membrane parameter estimates which were obtained by underestimating dendritic cable properties [17, 66]. The subsequent mathematical modeling of Rall and others can claim significant credit for contributing insight on this issue.

From equation (3) and our assumption, $I_{\text{ion}} = R_m^{-1} v$ (section 2), the equation for $v(x, t)$ in a constant diameter dendritic branch segment is

$$\lambda^2 \frac{\partial^2 v}{\partial x^2} = \tau_m \frac{\partial v}{\partial t} + v + R_m (I_{\text{syn}} - I_{\text{app}}). \qquad (26)$$

Here λ, the electrotonic length constant for the dendritic branch, is given by

$$\lambda = \left[R_m d / (4 R_i) \right]^{1/2}. \qquad (27)$$

It is proportional to the square root of the branch diameter d. The membrane time constant τ_m equals $R_m C_m$ and is independent of d. When synaptic input, for example in the excitatory case, is modeled as a conductance change $G_\varepsilon(x, t)$ we have

$$I_{\text{syn}}(x, t) = G_\varepsilon(x, t)(v - V_\varepsilon) \qquad (28)$$

where V_ε is the excitatory equilibrium potential, a constant.

Note, equation (26) is not merely an excitable membrane description such as HH linearized around the rest state. For such a derivation, the intrinsic, ionic permeabilities of the membrane would still depend on time. Here, those permeabilities are taken to be constant. Without synaptic input, (26) is often referred to, for historical reasons, as the "cable equation for electrotonic potential."

A mathematical model for dendritic integration would include the numerous branches in the dendritic trees. For some treatments, the soma is approximated by a uniform membrane patch, a "lumped soma" (e.g., see [46, 47, 61, 64, 66]). For others, some initial segment of each dendritic trunk accounts for the cell body [60, 65, 68]. We follow the latter approach.

To determine the potential distribution throughout a dendritic neuron model we must fully specify a mathematical problem. With given synaptic and/or externally applied input, a solution must satisfy (26) in each branch, in addition to certain initial and boundary conditions. It is reasonable to assume that no current will flow out the tiny dendritic terminals, i.e., their distal tips. This insulated end boundary condition is $\partial v / \partial x = 0$. At a dendritic branching point, the membrane potential is continuous and axial current is conserved. This latter requirement is a matching condition for the derivatives $\partial v / \partial x$ in the various segments which meet at the branching point.

In general, the problem which we have formulated is not analytically solvable. However, for a certain class of dendritic trees, the matching conditions at the branching points can be easily satisfied. Rall [60] realized this and exploited it to greatly simplify the mathematics and expose the qualitative and quantitative features of the phenomenology. In the next section we will describe Rall's simplification.

Approaches which treat arbitrary branching geometries generally require extensive computing. Recently, Barrett and Crill [4, 5] used numerical transform methods to calculate solutions corresponding to input at single locations. Their dendritic configuration was a reconstruction of a cat spinal motoneuron from which they had recorded experimentally and subsequently dissected. Transform solutions have also been obtained by Butz and Cowan [6] for arbitrary branching geometries. Time domain expressions for their solutions generally require numerical inversion.

4.2. Idealized branching. It is convenient to introduce a dimensionless time variable $T = t / \tau_m$ and a dimensionless distance variable $X(x)$. Electrotonic distance X is measured relative to distance from the "soma," the junction of all the dendritic trunks. For a given dendritic location, the physical distance x to the soma is the sum of the distance increments Δx_j for the different branches in the path to the soma. The electrotonic distance $X(x)$ is the sum of the Δx_j each divided by the electrotonic length constant λ_j for that branch:

$$X(x) = \int_0^x \lambda^{-1}(s) \, ds. \tag{29}$$

Thus in a given branch

$$dX/dx = 1/\lambda \tag{30}$$

and this quantity, according to (27), is proportional to $1/\sqrt{d}$. Now, let $V(X, T)$ denote membrane potential in terms of X and T. The cable equation (26) is then written as

$$\frac{\partial^2 V}{\partial X^2} = \frac{\partial V}{\partial T} + V + I_{\text{in}}(X, T) \tag{31}$$

where $I_{\text{in}}(X, T)/R_m$ denotes the combined input $I_{\text{syn}} - I_{\text{app}}$ expressed in terms of X and T.

To motivate the idealized branching criterion we consider the simple tree illustrated in Figure 15. Here input is applied to the parent branch at $X = 0$. The n identical siblings which emanate from the parent at $X = X_1$ are of length $L - X_1$. By symmetry, membrane potential is distributed identically in each of the siblings. At the branching point X_1, the axial current entering from the parent i_p is equal to n times that which is leaving in each sibling i_s; $i_p = ni_s$. According to equation (1) the axial current in a branch is proportional to the voltage gradient $\partial v/\partial x$ times d^2. Thus from (1) and (30) it follows that, in terms of electrotonic distance, the matching condition at X_1 is expressed as

$$\left.\frac{\partial V}{\partial X}\right|_{X = X_1^-} = n\left(\frac{d_s}{d_p}\right)^{3/2} \left.\frac{\partial V}{\partial X}\right|_{X = X_1^+}.$$

In general, this means that $\partial V/\partial X$ will experience a jump discontinuity at X_1. Suppose however the factor in parentheses is equal to $1/n$. Then, the derivative of V is continuous at X_1. In this case, the solution to the simple problem is thus obtained by solving (31) for $0 < X < L$ with $I_{\text{in}} = 0$; explicit boundary conditions are imposed only at $X = 0$ and $X = L$. Thus for branching which satisfies the criterion

$$n\, d_s^{3/2} = d_p^{3/2} \tag{32}$$

and for electrotonically symmetric input configurations, matching at the branching points is trivially accomplished.

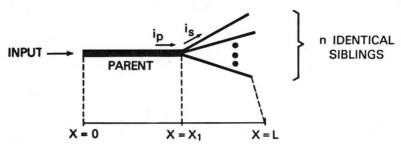

FIG. 15. Simple configuration used to illustrate consequences of idealized dendritic branching. From a parent branch emerge *n* identical sibling branches. Lower scale indicates electrotonic distance.

The fundamental notion of this simple example was described by Rall [60] for more general branching patterns in dendritic trees. For a class of problems, the branching geometry of a dendritic tree mathematically reduces to a single unbranched "equivalent cylinder." The principal assumptions which allow this reduction are:

(i) for each X, the sum of the branch diameters each raised to the 3/2-power is equal to $d_0^{3/2}$, where d_0 is the trunk diameter;

(ii) the dendritic terminals are equidistant, in terms of X, from the soma;

(iii) inputs, initial conditions, and boundary conditions are "electrotonically" symmetric.

Anatomical measurements reported by some investigators [51, 61] indicate that the 3/2-branching criterion is satisfied to a first approximation in dendritic trees of cat motoneurons. And while other investigators [4] have found that $\Sigma_j d_j^{3/2}(X)$ decreases with distance from the soma, the decrease is attributed largely to tapering of dendritic trunks. In this regard, Rall's theory [34, 60] can accommodate some forms of taper.

A sample problem. Consider the neuron model shown in Figure 16 (left). There are six identical trees each of length L and with two orders of 3/2-branching. For this case of symmetric bifurcations, branch diameter decreases by about 37% at each successive generation. A sample problem is: determine the response to a transient excitatory synaptic bombardment which is restricted to all dendritic locations in the distance interval $(L/3, 2L/3)$ and which is uniform in that interval. Suppose the neuron is initially at rest and its terminals are insulated. By symmetry we need consider only one of the trees. For this problem, we may apply the equivalent cylinder reduction; see the right side of Figure 16. The transient response is obtained by solving (31) for a single cylinder, $0 < X < L$, and for $T > 0$ with the input

$$I_{in}(X, T) \begin{aligned} &= \mathcal{E}(T)(V - V_\varepsilon), \quad \text{for } L/3 < X < 2L/3, \\ &= 0, \qquad\qquad\quad \text{otherwise,} \end{aligned}$$

where $\mathcal{E}(T) = R_m G_\varepsilon(\tau_m T)$. The initial and boundary conditions are $V(X, 0) = 0$ and

$$\frac{\partial V}{\partial X}(0, T) = 0, \quad \frac{\partial V}{\partial X}(L, T) = 0. \tag{33}$$

The boundary condition at $X = 0$ results from symmetry. It is important to realize that, because the equivalent cylinder concept involves a class of trees, the solution to this sample problem has application beyond the configuration of Figure 16.

NEURON MODEL

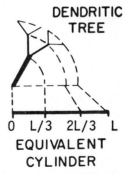

DENDRITIC TREE

0   L/3 2L/3   L
EQUIVALENT CYLINDER

FIG. 16. Idealized neuron model (left) composed of six identical dendritic trees with two orders of branching. The point of common origin is regarded as the neuron soma. On the right, a dendritic tree is related to its unbranched equivalent cylinder. For the sample problem (see text) each branch length is $L/3$.

4.3. Applications of a dendritic neuron model. *EPSP shapes.*
Calculated and analytical solutions to problems similar to the
sample one above illustrate several qualitative features of dendritic
integration in the neuron model. They show how the theoretical
soma potential depends upon input time course and location.
Generally, for a given input time course, more distant inputs result
in more slowly-varying, broader temporal waveforms at the soma
while proximal inputs produce sharper transients. When a given
input is applied successively to different locations, the spatio-temporal pattern could determine whether or not a cell might respond
with an action potential. As an example, Rall [61] compared a
centrifugal pattern $\mathcal{E}(t)$ which moved from the dendritic tips
toward the soma, with a centripetal pattern; the latter resulted in a
substantially larger amplitude soma potential. Similar calculations
illustrated the effects of inhibition, spatial summation (result of
simultaneous inputs to different locations), and temporal summation (result of time sequenced inputs to the same location).

Calculated EPSP's have also been quantitatively described with
shape indices such as time-to-peak and half-width [45, 62]. The
range of computed shape indices has accounted for experimental
ones obtained from cat motoneurons in various laboratories [43,
44, 62]. The associated range of input locations implicates dendritic activity which is widely distributed over the motoneuron
surface and not merely restricted to soma locations. The comparisons also reveal that significant soma depolarization could result
even for inputs located at dendritic sites of considerable distance
from the soma.

Neuronal parameter estimates. To determine estimates for electri-
cal-geometrical parameters, e.g., τ_m and L, of a dendritic neuron
we might perform the following experiment. Inject a brief current
into the cell body with a microelectrode. When the current is
turned off, we will have established a potential distribution in the
cell. Take it as an initial condition $V_0(X)$. Suppose there are no
other significant inputs to the cell. To describe the decay back to
rest of the potential distribution throughout the cell we could
employ the equivalent cylinder concept. This means solving (31)
with $I_{in} = 0$ subject to the boundary conditions (33) and the initial

condition $V(X, 0) = V_0(X)$. This problem can be solved explicitly, e.g., by separation of variables and eigenfunction expansions [64]. The solution is

$$V(X, T) = \sum_{n=0}^{\infty} c_n \cos(n\pi X/L) \exp\left[-\left(1 + \left(\frac{n\pi}{L}\right)^2\right)T\right] \quad (34)$$

where the c_n are determined by $V_0(X)$.

From the representation (34), the large time behavior of the potential recorded at the soma is

$$V(0, T) \sim c_0 \exp(-T) + c_1 \exp\left[-\left(1 + \frac{\pi^2}{L^2}\right)T\right], \quad T \gg 1.$$

$$(35)$$

Since $T = t/\tau_m$, an estimate for τ_m can be obtained from the ultimate decay of the membrane potential. For motoneurons, estimated values of τ_m are about 5 msec [44, 51]. Occasionally, the electrophysiological data will permit one to estimate L based on the decay rate of $V(0, T) - c_0 \exp(-T)$. Estimated values for L lie in the range of 1–2 [51, 54]. Detailed anatomical measurements (when combined with estimates for R_m and R_i) yield values which agree with this range [4, 51].

The above estimation technique is due to Rall [64]. It would also apply of course to the decay of $V(X, T)$ after a brief burst of synaptic activity. Techniques based upon other analytical representations of the solution for brief current injection have been described by Jack and Redman [45, 46]. Their recipes when applied to data from cat motoneurons yield estimates similar to those reported here.

Response for input to single branch locations. The calculations of theoretical EPSP's which were described above were performed by treating a dendritic tree as a single unbranched equivalent cylinder. Input to only a single branch location was not explicitly considered. This latter case however can be treated analytically under the 3/2-branching assumption. The transient response $K(X, T; X_{in})$ throughout the dendritic neuron model to a pulse of current (δ-function) at a single branch site X_{in} was determined explicitly by Rinzel and Rall [68]. They obtained their results by exploiting the principle of superposition in this linear system and, to simplify their exposition, they used Laplace transforms. With this fundamental solution, one can determine the response to an arbitrary current injection $I_{in}(T)$ at X_{in} according to the convolution formula

$$V(X, T) = \int_0^T K(X, T - s; X_{in})I_{in}(s)ds. \qquad (36)$$

In (36), the initial state is taken to be $V(X, 0) = 0$.

As an example, Rinzel and Rall calculated the response at various locations in the neuron model for a brief excitatory input current to a single branch terminal, $X_{in} = L$. The results are illustrated in Figure 17. The neuron model is shown upper right. There are six trees with three orders of symmetric branching. Here $L = 1.0$ and the branch lengths are identical. The input time course, shown upper left, corresponds to an EPSP shape in the range of those seen experimentally. Normalized voltage transients are plotted on a semi-log scale for BI(branch input), its ancestors: P(parent), GP(grandparent), GGP(great-grandparent), and the soma. These dendritic transients illustrate the attenuation and broadening characteristics of passive membrane behavior. The ratio of the peak amplitude at the input terminal to the attenuated peak response at the soma is about 235 for this case.

Such transient solutions can also be used to compare, for different X_{in}, potentials at the input sites and at the soma. For example, suppose the same input current as in Figure 17 is applied to the soma. The peak amplitude at the soma in this case is about seven times that of Figure 17 for input to a terminal. We see that, although the former case exhibited severe attenuation to the soma, the soma peak for the distal input is not orders of magnitude insignificant relative to that for the somatic input. The reason is that, because of different input impedances, the amplitude at a branch input terminal is large relative to that at a soma input location (see also, [5, 67]). Theoretical effectiveness comparisons have also been made on the basis of charge delivered to the soma [5, 43, 68].

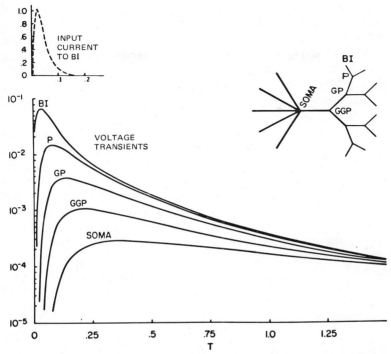

FIG. 17. Semi-log plots of scaled membrane potential versus T at several locations in the neuron model for transient current injected into the terminal of one branch. The normalized input current is shown upper left. BI designates the input branch terminal while P, GP, and GGP designate the parent, grandparent, and great grandparent nodes, respectively, along the mainline from BI to the soma. The neuron model shown upper right has six trees with three orders of branching, $L = 1.0$, and $\Delta X = .25$ (branch length).

As a special case, Rall and Rinzel [65] applied their results to the time-independent case of steady current injected to a single dendritic branch. This application produced analytical expressions for the steady attenuation of potential from X_{in} to the cell body and for the input resistance R_{in} at X_{in}. These results show for the neuron model how R_{in} depends on the number of dendritic trees and the number of orders of branching for a given X_{in}.

For such time independent problems, I can briefly outline how we applied superposition methods to obtain solutions; for details, see [65]. The transform solutions for the transient case are obtained in the same way. Consider the neuron model in Figure 16. It has six trees with two orders of symmetric bifurcations. The branch lengths are all $L/3$. Suppose a steady current I is applied to just one terminal; the other terminals are insulated. We obtain the solution as a sum of simpler component solutions as schematized in Figure 18. In A-C the trees are shown as their equivalent cylinders; branching is not treated explicitly at this stage. Input to just one tree, as shown in C, corresponds to the combination of A, $I/6$ to each of the trees, with B, $5I/6$ to the input tree and $-I/6$ to each of the others. To now treat the branching, we proceed from C; only the branching in the input tree matters. The configuration in C is redrawn in D with the input I equally divided among the branch terminals; the second order branches are still retained as their equivalent cylinders. The superposition of D with E leads to F. Note that symmetry implies zero potential for dendritic regions which are shown dashed. The second stage of branching is handled similarly. First, F is redrawn in G. The superposition of H with G yields J, the configuration we seek. The solution, for the input branch, is the sum of solutions to the component problems A, B, E, and H:

$$
V(X, T) = \left(\frac{6R_N I}{\coth L} \right) \left[\frac{1}{6} \frac{\cosh X}{\sinh L} + \frac{5}{6} \frac{\sinh X}{\cosh L} \right.
$$

$$
\left. + \frac{\sinh(X - L/3)}{\cosh(2L/3)} + 2\frac{\sinh(X - 2L/3)}{\cosh(L/3)} \right],
$$

where R_N is the input resistance of the neuron, seen at the soma.

A by-product of this superposition method is an explicit demonstration that the response at the soma is independent of the way I_{in} is distributed among the branch locations at X_{in}.

Neuron models with dendritic integration and axonal signaling. Finally I would like to briefly mention two applications in which

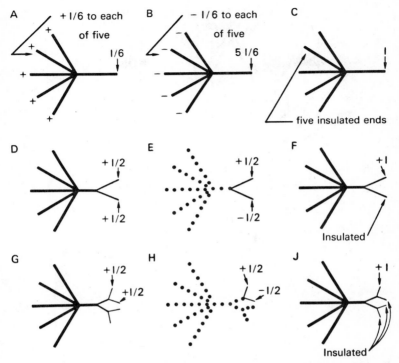

FIG. 18. Schematic diagrams which illustrate superposition technique to obtain time independent solution for steady current applied to one branch terminal. A-C are for input to all terminals of just one tree. D-H and J consider branching details of input tree.

a dendritic model was extended to include a soma and a short segment of axon with excitable membrane properties.

Dodge and Cooley [18] formulated a computational model of a dendritic motoneuron with (myelinated) axon. By matching voltage clamp data from motoneurons, they adapted the HH description of ionic current flows to model membrane behavior in the excitable regions. A schematic representation of their motoneuron and theoretical cable model is given in Figure 19. The dendritic structure is represented as an equivalent cylinder. Darker shading in the schematic cable indicates more "highly excitable" mem-

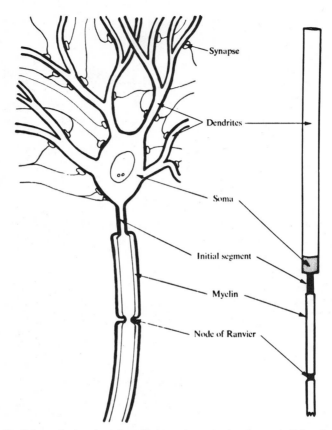

FIG. 19. Schematic drawing of a spinal motoneuron, showing a small fraction of all synaptic endings, and a representation of its electrical characteristics by a nonuniform cable. (From Dodge and Cooley [16], copyright 1973 by International Business Machines Corporation.)

brane behavior. With their model they could account for various experimental data which included action potential recordings from the soma. They concluded that a satisfactory match with the data required passive dendritic membrane. In further support of experimental interpretation [19, 20] they conclude that, relative to the axon's initial segment, soma membrane has a lower density of sodium conductance ($\bar{g}_{Na}$) and an intrinsically higher "voltage threshold" (modeled by translations of the rate curves in Figure 3). Based upon the agreement of the theoretical and experimental evidence for non-uniform distribution of excitability characteristics over the neuron membrane, they suggest a mechanism for its development and maintenance.

In a different application, Rall and Shepherd [63] used neuron models with passive and excitable membrane behavior to study the interactions between two populations of cells. Their motivation was to account for extracellular recordings obtained in the olfactory bulb of the rabbit. Corresponding to the experimental setup, they assumed that the two cell populations were distributed symmetrically and uniformly in a spherical arrangement. One population received direct synchronous input from impulses which were excited by an external stimulus at some location on their axons and which then propagated "back" toward the cell bodies and dendritic trees (this is called antidromic stimulation). Because of the symmetry and synchrony it was sufficient to analyze a representative portion of the spherical arrangement and thereby reduce the computational effort. To calculate dendritic potentials, they used the equivalent cylinder reduction. For excitable membrane behavior they used a three variable model (which was also used in [34]). From membrane potential and current distributions they determined theoretical extracellular potentials. To adequately match their calculated extracellular potentials with the experimental ones, Rall and Shepherd postulated that the two populations interacted by means of reciprocal synapses between their respective dendrites. The notions of dendro-dendritic synapses and reciprocal intercommunication are not included in the classical structure-function description of most physiologists. It is a tribute to the modeling here that anatomists subsequently discovered such connections in the olfactory bulb and other networks. This structural arrangement, particularly for sensory processing, suggests a mechanism for lateral inhibition.

5. SUMMARY

We have focused on two aspects of neuroelectric signaling at the cellular level and have discussed mathematical models for them. We have done so in the context of a classical, but not universal, structure-function description for neurons. In this view, an axon is for nerve impulse propagation over long distances. Its membrane is excitable and its dynamic behavior is described by a nonlinear parabolic partial differential equation. The axonal geometry is relatively simple. For dendritic integration, PSP's are transmitted over relatively short distances to the cell body. For some cells passive membrane behavior seems a reasonable assumption and, for them, membrane potential satisfies a linear equation. The geometry includes extensive dendritic branching.

For axonal signaling, we outlined first the elegant construction of the classical Hodgkin-Huxley theory based on voltage clamp data. From that data with its different time scales, we derived the simpler two-variable FitzHugh-Nagumo model which has an N-shaped nonlinearity. For this conceptual model, some qualitative features of excitability in the space-clamped case were illustrated. Then for steady impulse propagation we exploited the simplicity of a piecewise linear FitzHugh-Nagumo equation to obtain the fast and slow solitary pulse and the one-parameter family of periodic traveling waves. Explicit stability analysis (temporal and spatial) of these solutions was discussed. Results for this model equation suggest dispersive aspects of signaling and qualitative stability properties for a class of nerve conduction equations. To illustrate some threshold phenomena for impulse initiation we described the strength-duration curve for the solitary pulse case and the frequency-current curve for repetitive firing. We presented two examples of axonal structural variations, the noded axon and an axon with a step change in diameter, and discussed some consequences for nerve signaling.

For dendritic integration, we formulated a mathematical model based upon a linear cable equation for membrane potential in each dendritic branch. We described how, under certain symmetry and branching conditions, Rall simplifies the general formulation by treating a dendritic tree of a certain class as a single un-branched "equivalent cylinder". Various applications of the Rall model reveal qualitative and quantitative features of integration and transmission of dendritic membrane potential transients. For synaptic activity, the response at the soma reflects the spatio-temporal pattern of that activity. The temporal decay of the soma response after a brief input can be used to estimate neuronal parameters. For trees of the equivalent cylinder class, analytic solutions are obtained for input to just one branch site. These solutions have been used to illustrate attenuation from branch input sites to the soma, to compare the effectiveness of inputs at different sites, and to get expressions for input resistance at branch sites. For motoneurons, the model has been used to account for various experimental data, EPSP and action potential shapes, and to implicate the functional importance of dendritic synaptic activity. Use of the model to interpret data obtained from a rabbit's olfactory system, led to the understanding of an expanded role for dendritic signaling in this system.

Acknowledgment. I thank R. FitzHugh and W. Rall for their helpful comments on the manuscript.

REFERENCES

1. W. J. Adelman and R. FitzHugh, "Solutions of the Hodgkin-Huxley equations modified for potassium accumulation in a periaxonal space," *Federation Proc.*, **34** (1975), 1322–1329.

2. D. G. Aronson and H. F. Weinberger, "Nonlinear diffusion in population genetics, combustion, and nerve propagation," in *Partial Differential Equations and Related Topics*, Lecture Notes in Mathematics, vol. 446, Springer-Verlag, New York, 1975.

3. Y. I. Arshavskii, M. R. Berkinblit, and V. L. Dunin-Barkovskii, "Propagation of pulses in a ring of excitable tissues," *Biophysics*, **10** (1965), 1160–1166.

4. J. N. Barrett and W. E. Crill, "Specific membrane properties of cat motoneurones," *J. Physiol. (Lond.)*, **239** (1974), 301–324.

5. ———, "Influence of dendritic location and membrane properties on the effectiveness of synapses on cat motoneurones," *J. Physiol. (Lond.)*, **239** (1974), 325–345.

6. E. G. Butz and J. D. Cowan, "Transient potentials in dendritic systems of arbitrary geometry," *Biophys. J.*, **14** (1974), 661–689.

7. G. A. Carpenter, "A geometric approach to singular perturbation problems with applications to nerve impulse equations," *J. Differential Equations*, **23** (1977), 335–367.

7a. ———, "Periodic solutions of nerve impulse equations," *J. Math. Anal. Appl.*, **58** (1977), 152–173.

8. R. H. Casten, H. Cohen, and P. Lagerstrom, "Perturbation analysis of an approximation to Hodgkin-Huxley theory," *Quart. Appl. Math.*, **32** (1975), 365–402.

9. H. Cohen, "Nonlinear diffusion problems," in *Studies in Applied Mathematics* (A. H. Taub, ed.), Prentice-Hall, Englewood Cliffs, 1971, 27–64.

10. ———, "Mathematical developments in Hodgkin-Huxley theory and its approximation," in *Lectures on Mathematics in the Life Sciences*, vol. 8 (S. A. Levin, ed.), Amer. Math. Soc., Providence, 1975.

11. K. S. Cole, *Membranes, Ions and Impulses*, University of California Press, Berkeley, California, 1968.

12. K. S. Cole, H. A. Antosiewicz, and P. Rabinowitz, "Automatic computation of nerve excitation," *J. SIAM*, **3** (1955), 153–172.

13. K. S. Cole, R. Guttman, and F. Bezanilla, "Nerve membrane excitation without threshold," *Proc. Nat. Acad. Sci.*, **65** (1970), 884–891.

14. C. C. Conley, "Traveling wave solutions of nonlinear diffusion equations," in Proc. of Conf.: *Structural Stability, Catastrophe Theory and Their Applications in the Sciences* (P. F. Hilton, ed.), Springer-Verlag, New York, 1975.

15. J. Cooley, F. Dodge, and H. Cohen, "Digital computer solutions for excitable membrane models," *J. Cell. Comp. Physiol.*, **66**, Supp. 2 (1965), 99–109.

16. J. W. Cooley and F. A. Dodge, "Digital computer solutions for excitation and propagation of the nerve impulse," *Biophys. J.*, **6** (1966), 583–599.

17. J. S. Coombs, J. C. Eccles, and P. Fatt, "The electrical properties of the motoneurone membrane," *J. Physiol. (Lond.)*, **130** (1955), 291–325.

18. F. A. Dodge and J. W. Cooley, "Action potential of the motoneuron," *IBM J. Res. Devel.*, **17** (1973), 219–229.

19. J. C. Eccles, *The Physiology of Nerve Cells*, The Johns Hopkins Press, Baltimore, Maryland, 1957.

20. ———, *The Physiology of Synapses*, Academic Press, New York, 1964.

21. J. W. Evans, "Nerve axon equations: I. Linear approximations," *Indiana Univ. Math. J.*, **21** (1972), 877–885.

22. ———, "Nerve axon equations: II. Stability at rest," *Indiana Univ. Math. J.*, **22** (1972), 75–90.

23. ———, "Nerve axon equations: III. Stability of the nerve impulse," *Indiana Univ. Math. J.*, **22** (1972), 577–593.

24. ———, "Nerve axon equations: IV: The stable and the unstable impulse," *Indiana Univ. Math. J.*, **24** (1975), 1169–.

24a. J. W. Evans and J. Feroe, "Local stability theory of the nerve impulse," *Math. Bios.* (to appear).

25. R. FitzHugh and H. A. Antosiewicz, "Automatic computation of nerve excitation—detailed corrections and additions," *J. SIAM*, **7** (1959), 447–458.

26. R. FitzHugh, "Thresholds and plateaus in the Hodgkin-Huxley nerve equations," *J. Gen. Physiol.*, **43** (1960), 867–896.

27. ———, "Impulses and physiological states in models of nerve membrane," *Biophys. J.*, **1** (1961), 445–466.

28. ———, "Computation of impulse initiation and saltatory conduction in a myelinated nerve fiber," *Biophys. J.*, **2** (1962), 11–21.

29. ———, "Mathematical models of excitation and propagation in nerve," in *Biological Engineering* (H. P. Schwan, ed.), McGraw-Hill, New York, 1969, 1–85.

30. ———, "Dimensional analysis of nerve models," *J. Theoret. Biol.*, **40** (1973), 517–541.

31. S. V. Fomin, "Some aspects of information processing in the nervous system," Proc. IV Interntl. Biophys. Cong., Moscow, 1972.

32. H. A. Fozzard and M. Schoenberg, "Strength-duration curves in cardiac Purkinje fibres: effects of liminal length and charge distribution," *J. Physiol. (Lond.)*, **226** (1972), 593–618.

33. L. Goldman and J. S. Albus, "Computation of impulse conduction in myelinated fibres; theoretical basis of the velocity-diameter relation," *Biophysical J.*, **8** (1968), 596–607.

34. S. S. Goldstein and W. Rall, "Changes of action potential shape and velocity for changing core conductor geometry," *Biophys. J.*, **14** (1974), 731–757.

35. S. P. Hastings, "The existence of periodic solutions to Nagumo's equation," *Quart. J. Math.*, Oxford, **25** (1974), 369–378.

36. ———, "The existence of homoclinic and periodic orbits for the FitzHugh-Nagumo equations," *Quart. J. Math.*, Oxford, **27** (1976), 123–134.

37. ———, "Some mathematical problems from neurobiology," *Amer. Math. Monthly*, **82** (1975), 881–894.

38. A. V. Hill, "Excitation and accommodation in nerve," *Proc. Roy. Soc. Lond. B*, **119** (1936), 305–355.

39. A. L. Hodgkin, *The Conduction of the Nerve Impulse*, Charles C. Thomas, Springfield, Illinois, 1964.

40. A. L. Hodgkin and A. F. Huxley, "A quantitative description of membrane current and its application to conduction and excitation in nerve," *J. Physiol. (Lond.)*, **117** (1952), 500–544.

41. P. J. Hunter, P. A. McNaughton and D. Noble, "Analytical models of propagation in excitable cells," *Prog. Biophys. Molec. Biol.*, **30** (1975).

42. A. F. Huxley, "Can a nerve propagate a subthreshold disturbance?" *J. Physiol. (Lond.)*, **148** (1959), 80–81P.

43. R. Iansek and S. J. Redman, "The amplitude, time course and charge of unitary excitatory post-synaptic potentials evoked in spinal motoneurone dendrites," *J. Physiol. (Lond.)*, **234** (1973), 665–688.

44. J. J. B. Jack, S. Miller, R. Porter, and S. J. Redman, "The time course of minimal excitatory post-synaptic potentials evoked in spinal motoneurones by group Ia afferent fibres," *J. Physiol. (Lond.)*, **215** (1971), 353–380.

45. J. J. B. Jack and S. J. Redman, "The propagation of transient potentials in some linear cable structures," *J. Physiol. (Lond.)*, **215** (1971), 283–320.

46. ———, "An electrical description of the motoneurone, and its application to the analysis of synaptic potentials," *J. Physiol. (Lond.)*, **215** (1971), 321–352.

47. J. J. B. Jack, D. Noble, and R. W. Tsien, *Electric Current Flow in Excitable Cells*, Clarendon Press, Oxford, 1975.

48. B. Katz, *Nerve, Muscle, and Synapse*, McGraw-Hill, New York, 1966.

49. B. I. Khodorov, *The Problem of Excitability*, Plenum Press, New York, 1974.

50. R. Llinás and C. Nicholson, "Electrophysiological properties of dendrites and somata in alligator Purkinje cells," *J. Neurophys.*, **34** (1971), 532–551.

51. H. D. Lux, P. Schubert, and G. W. Kreutzberg, "Direct matching of morphological and electrophysiological data in cat spinal motoneurons," in *Excitatory Synaptic Mechanisms* (P. Anderson and J. K. S. Jansen, eds.), Universitetsforlaget, Oslo, Norway, 1970, 189–198.

52. H. P. McKean, "Nagumo's equation," *Advances in Math.*, **4** (1970), 209–223.

53. J. S. Nagumo, S. Arimoto, and S. Yoshizawa, "An active pulse transmission line simulating nerve axon," *Proc. IRE.*, **50** (1962), 2061–2070.

54. P. G. Nelson and H. D. Lux, "Some electrical measurements of motoneuron parameters," *Biophys. J.*, **10** (1970), 55–73.

55. D. Noble, "Applications of Hodgkin-Huxley equations to excitable tissues," *Physiol. Rev.*, **46** (1966), 1–50.

56. D. Noble and R. B. Stein, "The threshold conditions for initiation of action potentials by excitable cells," *J. Physiol.*, **187** (1966), 129–162.

57. D. Noble, "The relation of Rushton's liminal length for excitation to the resting and active conductances of excitable cells," *J. Physiol.*, **226** (1972), 573–591.

58. F. Ramon, R. W. Joyner, and J. W. Moore, "Propagation of action potentials in inhomogeneous axon regions," *Federation Proc.*, **34** (1975), 1357–1363.

59. J. Rauch and J. Smoller, "Qualitative theory of the FitzHugh-Nagumo equations," *Advances in Math.* (to appear).

60. W. Rall, "Theory of physiological properties of dendrites," *Ann. N.Y. Acad. Sci.*, **96** (1962), 1071–1092.

61. ———, "Theoretical significance of dendritic trees for neuronal input-output relations," in *Neural Theory and Modeling* (R. F. Reiss, ed.), Stanford University Press, Stanford, California, 1964, 73–97.

62. W. Rall, R. E. Burke, T. G. Smith, P. G. Nelson, and K. Frank, "Dendritic location of synapses and possible mechanisms for the monosynaptic EPSP in motoneurons," *J. Neurophysiol.*, **30** (1967), 1169–1193.

63. W. Rall and G. M. Shepherd, "Theoretical reconstruction of field potentials and dendrodendritic synaptic interactions in olfactory bulb," *J. Neurophysiol.*, **31** (1968), 884–915.

64. W. Rall, "Time constants and electrotonic length of membrane cylinders and neurons," *Biophys. J.*, **9** (1969), 1483–1508.

65. W. Rall and J. Rinzel, "Branch input resistance and steady attenuation for input to one branch of dendritic neuron model," *Biophys. J.*, **13** (1973), 648–688.

66. W. Rall, "Core conductor theory and cable properties of nerve cells," in *The Nervous System, Vol. I, Cellular Biology of Neurons* (E. R. Kandel, ed.), *Am. Physiol. Soc.*, Washington, D. C., 1976.

67. S. J. Redman, "The attenuation of passively propagating dendritic potentials in a motoneurone cable model," *J. Physiol. (Lond.)*, **234** (1973), 637–664.

68. J. Rinzel and W. Rall, "Transient response in a dendritic neuron model for current injected at one branch," *Biophys. J.*, **14** (1974), 759–790.

69. J. Rinzel and J. B. Keller, "Traveling wave solutions of a nerve conduction equation," *Biophys. J.*, **13** (1973), 1313–1337.

70. J. Rinzel, "Spatial stability of traveling wave solutions of a nerve conduction equation," *Biophys. J.*, **15** (1975), 975–988.

71. ———, "Neutrally stable traveling wave solutions of nerve conduction equations," *J. Math. Biol.*, **2** (1975), 205–217.

72. ———, "Simple model equations for active nerve conduction and passive neuronal integration," in *Lectures on Mathematics in the Life Sciences*, vol. 8 (S. A. Levin, ed.), Amer. Math. Soc., Providence, 1975.

73. ———, "Nerve signaling and spatial stability of wave trains," in Proc. of Conf.: *Structural Stability, Catastrophe Theory and Their Applications in the Sciences* (P. J. Hilton, ed.), Springer-Verlag, New York, 1975.

73a. ———, "Repetitive activity and Hopf bifurcation under point-stimulation for a simple FitzHugh-Nagumo nerve conduction model," *J. Math. Biol.* (to appear).

74. W. A. H. Rushton, "Initiation of the propagated disturbance," *Proc. Roy. Soc. B*, **124** (1937), 210–243.

75. N. H. Sabah and R. A. Spangler, "Repetitive response of the Hodgkin-Huxley model for the squid giant axon," *J. Theoret. Biol.*, **29** (1970), 155–171.

76. D. H. Sattinger, "On the stability of waves of nonlinear parabolic systems," *Advances in Math.*, **22** (1976), 312–355.

77. A. C. Scott, "The electrophysics of a nerve fiber," *Rev. Modern Phys.*, **47** (1975), 487–533.

78. G. M. Shepherd, *The Synaptic Organization of the Brain*, Oxford University Press, New York, 1974.

79. R. B. Stein, "The frequency of nerve action potentials generated by applied currents," *Proc. Roy. Soc. Lond. B*, **167** (1967), 64–86.

80. C. F. Stevens, *Neurophysiology: A Primer*, John Wiley, New York, 1966.

81. W. C. Troy, "Oscillation phenomena in nerve conduction equations," Doctoral thesis, State University of New York at Buffalo, 1974.

82. A. T. Winfree, "Rotating solutions to reaction-diffusion equations in simply-connected media," in *SIAM-AMS Proc. 8: Mathematical Aspects of Chemical and Biochemical Problems and Quantum Chemistry* (D. S. Cohen, ed.), Amer. Math. Soc., Providence, 1975.

SOME ASPECTS OF THE 'EIGEN-BEHAVIOR' OF NEURAL NETS

J. D. Cowan and G. B. Ermentrout

0. INTRODUCTION

Neural nets are aggregates of nerve cells, the interactions of which generate patterns of electrochemical activity. Such patterns form a substrate by means of which animals interact with their environment. In this paper the dynamics of such activity is described in general terms, and a more detailed analysis is given of one of the simpler problems.

In what follows we characterize the activity generated in a block or slab of neural tissue comprising a very large number of closely packed and coupled nerve cells. Such a slab may be thought of as a portion of the cortex [G. Shepherd 1974], i.e., the outermost layers of nerve cells that make up the telencephalon or 'thought-brain' of all higher vertebrates, within which are located specialized areas such as the various sensory and motor cortexes, as well as nonspecific or association cortex, and the more primitive archicortex. The primate cortex, for example, comprises some 10^{10} nerve cells, and at least the same number of satellite cells, closely packed together into a slab some 2 or 3 mm thick, and with a surface area of about 8×10^4 mm^2, so that there are about 5×10^4 cells per mm^3, corresponding to a mean cellular separation of about 27 μm. Since cellular diameters vary from 10 to 30 μm, the cells are indeed close-packed, and their ancilliary structures, dendrites and axons, form a dense plexus or neuropil.

The junctions between cellular structures, axons, dendrites and cell bodies or somas, are termed *synapses*, and are the site of a complex sequence of electrochemical events that result in the transmission of cellular activity from presynaptic to postsynaptic sites. The larger cortical cells (pyramidal cells) have a total surface area of about 4×10^3 (μm)^2, and can therefore admit up to about 4×10^4 synaptic contacts, i.e., terminals from all cells located within a radial distance of about 250 μm, together with the terminals of peripheral, i.e., extracortical, axons. Thus in any cortical column of cross-sectional area 0.01 mm^2 there are about 10^3 cells, and up to 10^7 synaptic contacts between them. Such packing and contact densities are far in excess of those currently obtaining in modern computers, and point to perhaps the main difference between brains and computers, in that the anastomatic cortical net evidently supports a high level of simultaneous *cooperative* activity [R. L. Beurle 1956] as distinct from the essentially serial nature of contemporary computer systems.

In order to develop any model for such activity, some account of the details of cellular interconnection is required, as are the physiological properties of the cells. So far as the former is concerned, recent anatomical investigations of cortical microcircuitry [J. Lund *et al.* 1975] indicate that the cortex is very highly organized. The cell-types are found at specific locations, and even the interconnections between different cell types are spatially selective. J. Szentagothai [1973] has estimated that there may be as many as 50 different cell types distributed throughout the depths of the neocortex. Given that all these cell types are mutually interconnected in a highly specific and ordered fashion, there is evidently a very large measure of 'positional information' [L. Wolpert 1971] stored in the neocortex. More global investigations of the functional (physiological) organization [D. H. Hubel and T. N. Wiesel 1974 a, b] indicate that the cortex, at least portions of visual neocortex, displays a highly regular, almost crystalline organization, which provides at every cortical location, an array of cell types each capable of coding a specific stimulus parameter. Thus primary visual cortex, area 17, is known [J. Pettigrew 1977] to code locally, for such stimulus properties as visual direction, orientation, direction and velocity of movement, ocular dominance and binocular disparity, and to some extent size, shape, and texture. Hubel and Wiesel [1974a] have estimated that as much as 1 mm^2 of cross-sectional area of the cortex (i.e., 2-3 mm^3) is devoted to such local stimulus analysis. In terms of our previous analysis, this corresponds to some 10^5 cells, and about 10^9 synapses. In analyzing such a system, it is evident that a large number of different processes, each involving a different subset of the cortical net, must be considered, and ultimately pieced together to provide an adequate representation of the whole. In this paper we concentrate solely on the generalized individual process, without reference to the parameters such a process is presumed to encode. Instead we seek to answer the general question as to how any subset of the cortical net, comprising, say, 10^4 cells can actually encode in a stable and reliable fashion, stimulus parameters.

The knowledgeable reader will recognize this as an old and venerable problem in the theory of neural nets and automata, first posed by W. S. McCulloch and W. Pitts [1943] in their classic paper on the representation of events by neural nets. In the present paper we pose this problem once again not, however, with reference to the properties of finite state switching nets, but in terms of the properties of nonlinear fields or continua, which we introduce as a suitable representation of the activity seen in neural nets comprising large numbers of densely interconnected cells. The reader is referred to an earlier paper [J. D. Cowan 1974], for an adequate bibliography.

1. THE MATHEMATICAL MODEL

So far as is currently known, cortical neurons can be classified in two ways. They seem to be either excitatory or else inhibitory in their effects on other neurons [J. C. Eccles 1964]. Thus not only is the same chemical transmitter presumed to be released at all post synaptic sites, but the effect on other cells is presumed to be identical. Secondly, cells may make either short-range connections

with other cells, by means of axo-somatic, axo-dendritic, or dendro-dendritic synapses, or even by way of local electrotonic coupling [F. O. Schmitt *et al.* 1976]; or else they make long-range
connections by way of specially insulated (myelinated) axons. In
this paper we shall concentrate exclusively on the consequences of
the short-range interactions.

Individual nerve cells, like all cells, interact with their environments by way of their membranes. In principle for each cell
centered at the point **r**, the cellular response can be represented by
an appropriate state variable, the transmembrane potential $v(\mathbf{r}, t)$,
and a functional equation can be written for $v(\mathbf{r}, t)$, that takes
account of the cable properties of the membrane [W. Rall 1967],
its excitability [A. Hodgkin and A. Huxley 1952], and the various
modes of intercellular coupling. Given that the collective behavior
of some 10^4 cells is to be represented, it is obviously not feasible to
proceed in such a fashion. Instead of 'facing reality' [J. C. Eccles
1971], one has to abstract from it. The cells will therefore be
presumed to be 'space-clamped' so far as their individual membrane voltage is concerned, and to summate incoming excitation
both spatially and temporally in a linear time invariant fashion.
For a state-variable, we introduce the local spatial average of v,
the quantity

$$V(\mathbf{r}, t) = \int v(\mathbf{r} - \mathbf{r}', t)\rho(\mathbf{r}') \, d\mathbf{r}'$$

where $\rho(\mathbf{r})$ is the local packing density [see J. L. Feldman and J. D.
Cowan 1975, for details]. Then V follows the state equation,

$$L_i V_i(\mathbf{r}, t) = \psi_i(\mathbf{r}, t) \tag{1.1}$$

where L_i is a time-invariant operator that represents the membrane response, $\psi_i(\mathbf{r}, t)$ the presynaptic excitation, physically a
current, and where the index i refers to the ith cell-type. To
calculate ψ_i, the 'field' quantity, we proceed as follows. Let $a_j(\mathbf{r})$ be
the density of axonal fibers of the jth cell type present at **r**, and let
$d_i(\mathbf{r})$ be the corresponding dendritic fiber density of cells of the ith
type at **r**. Following A.M.Uttley [1955], we suppose that whenever
an axon and a dendrite grow to within a critical distance s, a

process of growth modification is initiated whereby a connection is formed. Then the mean number of connections between the axonal and dendritic systems is given by the convolution

$$\beta_{ij}(\mathbf{r}) = \frac{4s}{\pi} \int a_j(\mathbf{r}') \, d_i(\mathbf{r} - \mathbf{r}') \, d\mathbf{r}'$$

$$= \frac{4s}{\pi} (a_j \times d_i). \tag{1.2}$$

In similar fashion the mean number of axo-somatic connections β_{ik} is given by the expression

$$\beta_{ik}(\mathbf{r}) = \frac{4s}{\pi} a_k(\mathbf{r}) \tag{1.3}$$

and the mean number of dendro-dendritic connections by the convolution

$$\beta_{il}(\mathbf{r}) = \frac{4s}{\pi} (d_i \times d_l). \tag{1.4}$$

The field itself consists of currents, either in pulse-form when transmitted by axons at a propagation velocity v_j or more slowly varying when transmitted by dendrites or somas. In all such cases, if $\mathcal{I}_j(\mathbf{r}, t)$ is the appropriate current density, then

$$\psi_{ij}(\mathbf{r}, t) = \beta_{ij} \times \mathcal{I}_j(t - \tau_j), \tag{1.5}$$

where $\tau_j = |\mathbf{r} - \mathbf{r}'|/v_j$ for axonal transmission, and 0 otherwise.

It remains to determine the form of $\mathcal{I}_j(\mathbf{r}, t)$. In the case of current pulses transmitted by axons, the pulses result from the 'firing' of the cell in response to suprathreshold currents, an 'all-or-nothing' phenomenon [Eccles 1964]. Various statistical arguments can be given [J. D. Cowan 1970, H. R. Wilson and J. D. Cowan 1972, 1973, J. L. Feldman and J. D. Cowan 1975] all of which suggest that the relation between the mean current density $\mathcal{I}_j(\mathbf{r}, t)$ and the mean voltage $V_j(\mathbf{r}, t)$ is typified by a smooth sigmoidal curve. Thus in such a case

$$\mathcal{I}_j(\mathbf{r}, t) = \rho_j(\mathbf{r})\phi_j[V_j(\mathbf{r}, t)], \tag{1.6}$$

where $\rho_j(\mathbf{r})$ is the packing density of the jth cell type and where ϕ_j is as shown in Figure 1. ϕ_j may be thought of as a nonlinear or voltage controlled conductance. A slight modification of this argument suffices for dendro-dendritic synapses.

Finally the excitation produced by direct electrotonic coupling can be incorporated [V. Torre 1976], as

$$\mathcal{G}_j(\mathbf{r}, t) = \rho_j(\mathbf{r}) g_{ij}(V_j(\mathbf{r}, t) - V_i(\mathbf{r}, t)), \tag{1.7}$$

where g_{ij} is the effective conductivity of a single electrotonic junction.

On the presumption of linearity of the membrane response, equations (1.2) to (1.7) may be combined to give an expression for the field ψ_i, namely

$$\psi_i(\mathbf{r}, t) = \sum_j \varepsilon_j \beta_{ij} \times \rho_j \phi_j \big[V_j(\mathbf{r}, t') \big]$$

$$+ \sum_k \beta_{ik} \times \rho_k g_{ik}(V_k - V_i), \tag{1.8}$$

where the index j refers to all classes of synaptic interactions, and the index k to electrotonic coupling, and where ε equals $+1$ for excitation and -1 for inhibition. Equations (1.1) and (1.8) determine the spatio-temporal evolution of the 'field' of neural activity, represented by the mean voltage $V_i(\mathbf{r}, t)$.

The form of the membrane operator L_i can be derived as follows. We look first at the equation

$$L_i V_i(\mathbf{r}, t) = \delta(t) \tag{1.9}$$

defining the temporal Green's function $h_j = L_j^{-1}\delta(t)$, the response of the cell membrane to a current pulse delivering unit charge per millisecond. It follows that

$$V_i(\mathbf{r}, t) = \int_{-\infty}^{t} \psi_i(\mathbf{r}, t - t') h_i(t') \, dt'$$

$$= \psi_i \otimes h_i, \tag{1.10}$$

the temporal convolution of ψ_i and h_i. In general h_i consists of several components. The primary component has a time constant of a few milliseconds and represents the initial post-synaptic membrane response. Neglecting cable properties, it takes the form $C^{-1} \exp(-t/RC)$ where C is the membrane capacitance, R the membrane resistance, and the time constant $\tau_1 = RC$ is between 3 and 10 msec. In addition, following activation of the cell by excitation strong enough to build up the membrane potential to the presumed firing threshold v_i^{TH}, a process of 'afterhyperpolarisation' occurs [D. Kernell 1968], that effectively raises v_i^{TH}, so that the cell *adapts* or is fatigued. This is represented by the convolution $\phi_i[V_i] \otimes h_2$, where h_2 is another function of the form $\beta \exp(-t/\tau_2)$, with a time constant τ_2 of the order of tens of msec. In a similar fashion *accommodation* to very long lasting or slowly increasing stimuli can be incorporated by way of the convolution $V_i \otimes h_3$. The defining equation for $V_i(\mathbf{r}, t)$ thus takes the form

$$V_i(\mathbf{r}, t) = \psi_i \otimes h_1 - \phi_i[V_i] \otimes h_2 - V_i \otimes h_3. \quad (1.11)$$

It is to be expected that the field represented by equations (1.8) and (1.11) will exhibit very complicated dynamical structure. In what follows we will present a survey of some of the properties of only the simpler systems; we will not deal with electrotonic coupling nor with adaptation and accommodation, except in general terms. However, even the simpler systems, comprising non-adapting or non-accommodating excitatory and inhibitory cells, coupled synaptically, display a rich and interesting dynamical structure relevant to the representation problem discussed in §0.

2. SYNAPTICALLY COUPLED NETS

In order to provide the reader with some insight into the dynamics of neural nets we consider first the properties of a net comprising a single excitatory cell type, in which equations (1.8) and (1.11) reduce respectively, to

$$\psi_1(\mathbf{r}, t) = \beta_{11} \times \rho_1 \phi_1[V_1] + P_1, \quad (2.1)$$

$$V_1(\mathbf{r}, t) = \psi_1 \otimes h_1, \quad (2.2)$$

where $P_1(\mathbf{r}, t)$ is an external current applied to the net. Furthermore the net is assumed to be spatially homogeneous, so that the cellular packing density $\rho_1(\mathbf{r})$ is a constant ρ_1. Given the form of h_1 introduced in §1, these equations can be combined into the single non-linear integro-differential equation,

$$C\frac{\partial V_1}{\partial t} = -\frac{V_1}{R} + \beta_{11} \times \rho_1\phi_1[V_1] + P_1. \qquad (2.3)$$

As for the non-linearity $\phi_1[V_1]$, it will be assumed to be of the form

$$\phi[V] = \frac{q_0}{r+1}\left[1 + \exp(-\nu(V - \theta))\right]^{-1},$$

where q_0 is the mean charge delivered per unit impulse, r the absolute refractory period of the cells, measured in millisecs., ν a constant fixing the sensitivity to excitation of the population [see Wilson and Cowan] and θ the threshold mean voltage [see Feldman and Cowan 1975]. This choice of θ ensures that $\theta \to (r + 1)^{-1}$ as $V \to \infty$, so that if $r = 1$ ms, then ϕ, a measure of the proportion of cells becoming activated per unit time in the net, tends to 0.5 as V increases, in keeping with the intuitive notion that if all cells fire at their maximum rate of $1/r = 1000$ impulses/sec., then on the average only 50% of the cells will be active at any instant, since about 50% of the cells will be refractory then and unable to fire.

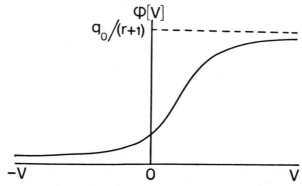

FIG. 1. 'Current-voltage' characteristic of a single homogeneous net.

In choosing a mathematical form for $\beta_{11}(\mathbf{r})$, the mean number of synaptic connections (of all classes of cells) with any given cell, from all cells a distance $(\mathbf{r})$ away, we have a great deal of latitude. In the present context we assume $\beta_{11}(\mathbf{r})$ to be either a rectangular function,

$$\beta_{11}(\mathbf{r}) = \begin{cases} b_{11}, & \text{for } |\mathbf{r}| \leq \sigma_{11}, \\ 0, & \text{otherwise}, \end{cases} \tag{2.4}$$

or a decaying exponential,

$$\beta_{11}(\mathbf{r}) = b_{11} \exp(-|\mathbf{r}|/\sigma_{11}), \tag{2.5}$$

whence eqn. (2.3) can be rewritten as

$$C \frac{\partial V_1}{\partial t} = -\frac{V_1}{R} + \frac{b_{11}\rho_1}{(r+1)} \int \exp(-|\mathbf{r} - \mathbf{r}'|/\sigma_{11}) \cdot q_0$$

$$\cdot \left[1 + \exp\left(-\nu(V_1(\mathbf{r}') - \theta_1(\mathbf{r}'))\right) \right]^{-1} \cdot d\mathbf{r}' + P_1. \tag{2.6}$$

This can be rewritten in the dimensionless form:

$$\frac{\partial X_1}{\partial T} = -X_1 + \varepsilon_{11} \int \exp(-|\mathbf{s} - \mathbf{s}'|)$$

$$\cdot \left[1 + \exp\left(-(X_1 - \theta_1)\right) \right]^{-1} d\mathbf{s}' + Y_1, \tag{2.7}$$

where $T = t/\tau$, $\tau = RC$, $X = \nu V$, $Y_1 = \nu R P_1$, and where the dimensionless coupling parameter ε_{11} is given by:

$$\varepsilon_{11} = \frac{b_{11}\rho_1\sigma_{11}\tau}{(r+1)} \cdot \frac{\nu q_0}{C}. \tag{2.8}$$

So far nothing has been assumed about the range of integration of the convolution $\beta \times \rho\phi$. Consider therefore a one-dimensional model $\mathbf{s} = x$, defined on a finite length $2L$ of excitatory neural tissue, with Neumann boundary conditions. Eqn. (2.7) specialises to

$$\frac{\partial X}{\partial T} = -X + \varepsilon \int_{-L}^{L} \exp(-|x - x'|)$$

$$\cdot \left[1 + \exp\left(-(X - \theta)\right) \right]^{-1} \cdot dx' + Y \tag{2.9}$$

with boundary conditions

$$\frac{\partial X}{\partial x}\bigg|_{x=L} = \frac{\partial X}{\partial x}\bigg|_{x=-L} = 0. \qquad (2.10)$$

We consider first the case when Y is such as to fix X at θ, whence $\partial X/\partial T = \partial X/\partial x = 0$, and

$$Y - \theta + \frac{\varepsilon}{2} \int_{-L}^{L} \exp(-|x - x'|) \, dx' = 0,$$

i.e.,

$$Y - \theta + \varepsilon[1 - \exp(-L) \cosh x] = 0. \qquad (2.11)$$

Thus as $L \to \infty$, $Y \to \theta - \varepsilon$, which is also independent of X. Since we will be considering the tissue only in regions near the origin $\cosh X$ will also be small. These arguments motivate us to replace the finite range, non-singular equation (2.9), which has a discrete spectrum, with an easier singular equation,

$$\frac{\partial X}{\partial T} = -X + \varepsilon \int_{-\infty}^{\infty} \exp(-|x - x'|)$$

$$\cdot \left[1 + \exp(-(X - \theta))\right]^{-1} dx' + Y, \qquad (2.12)$$

but still employ as trial solutions, those corresponding to the discrete spectrum. This procedure is essentially that followed by J. J. Kozak, S. A. Rice, and J. D. Weeks [1971] in their attempts to model phase transitions in fluids, and in fact equation (2.12) is a singular Hammerstein equation closely related to the equation introduced by these authors.

Properties of the excitatory net

We consider first the case of all-to-all connectivity, in which $\sigma \gg L$, so that X and Y are independent of position. Equation (2.8) then gives $\varepsilon = b\rho\tau\nu q_0/(r + 1)C$, where b is the mean number of synaptic connections from all cells in the net on any given cell, and equation (2.12) reduces to the nonlinear differential equation,

$$\frac{\partial X}{\partial T} = -X + 2\varepsilon\left[1 + \exp(-(X - \theta))\right]^{-1} + Y. \quad (2.13)$$

Let $U = X - \theta$, $W = Y - \theta$, then equation (2.13) reduces to

$$\frac{\partial U}{\partial T} = -U + 2\varepsilon\left[1 + e^{-U}\right]^{-1} + W, \quad (2.14)$$

so that $U_T = 0$ when $W = -\varepsilon$, as expected. This equation is first order in time, so that Liapunov's criterion for stability [see A. A. Andronov, A.A. Witt, and S.E. Khaikin 1966] can be immediately applied. This states that if $x_T = f(x, \lambda)$, then the system is stable at an equilibrium $\bar{x}$, if $f_x(\bar{x}, \lambda) < 0$ and unstable if $f_x(\bar{x}, \lambda) > 0$. Thus the equilibria of equation (2.14), defined by the equation

$$f(U; \varepsilon, W) = W + 2\varepsilon\left[1 + e^{-U}\right]^{-1} - U = 0, \quad (2.15)$$

are unstable or stable depending upon the sign of f_U. A plot of this function is shown in Figure 2. It will be seen that there are at most 2 stable equilibria, $\bar{U}_1$, and $\bar{U}_2$ defined (locally) by the condition

$$f_U(\bar{U}; \varepsilon) = 2\varepsilon e^{-\bar{U}} \cdot \left[1 + e^{-\bar{U}}\right]^2 - 1 < 0. \quad (2.16)$$

It is apparent that (globally) there is a critical value of ε, in this case $\varepsilon_c = 2$, such that for values of $\varepsilon < \varepsilon_c$ only one stable equilibrium exists, and for $\varepsilon > \varepsilon_c$, two stable equilibria exist.

These phenomena can be represented in a more global and perhaps illuminating fashion, as follows. We note that since equation (2.14) is first order it can be written as a *gradient system*, [M. W. Hirsch and S. Smale 1974], of the form

$$U_T = -\text{grad}_U V(U; \varepsilon, W), \quad (2.17)$$

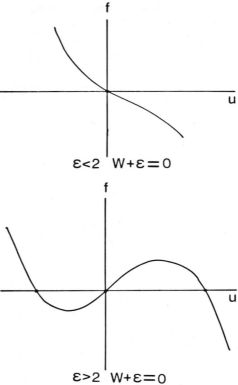

FIG. 2. See text for details.

where the potential function V takes the form

$$V(U; \varepsilon, W) = \frac{U^2}{2} - 2\varepsilon \left[U + \ln \frac{1}{2}(1 + e^{-U}) \right] - WU. \quad (2.18)$$

Liapunov's test for stability now amounts to a test for the existence of maxima or minima of the function V, i.e., by computing V_{UU}, at equilibria. Thus when $W + \varepsilon = 0$, $V = 0$, $V_U = 0$, and

$$V_{UU} = 1 - \varepsilon/2. \quad (2.19)$$

Therefore $V(0; \varepsilon, -\varepsilon)$ is an isolated minimum of V provided $\varepsilon < 2$. In fact $V(U; \varepsilon < 2, -\varepsilon)$ is a strict *Liapunov function* [see Hirsch and Smale 1974], so that in this case $\bar{U} = 0$ is an isolated asymptotically stable equilibrium. Conversely for $\varepsilon > 2$, $V(0; \varepsilon > 2, -\varepsilon)$ is a local maximum of $V(U; \varepsilon > 2, -\varepsilon)$, and hence $U = 0$ is now unstable. Figure 3a shows such behavior in cases when $\varepsilon > 2$. Even though W is fixed so that $U_T = U = 0$, such an equilibrium is unstable, and a *bifurcation* to either of the two stable equilibria $\bar{U}_+ > 0$ and $\bar{U}_- < 0$ will occur. Similar results obtain for cases when $W \neq -\varepsilon$. The critical point now occurs when

$$\varepsilon_c = \frac{1}{2} e^{\bar{U}} \big[1 + e^{-\bar{U}} \big]^2, \qquad (2.20)$$

where $\bar{U}$ is a solution of equation (2.15). Figure 3b shows the locus of such points in the plane $(W + \varepsilon, \frac{\varepsilon}{2} - 1)$.

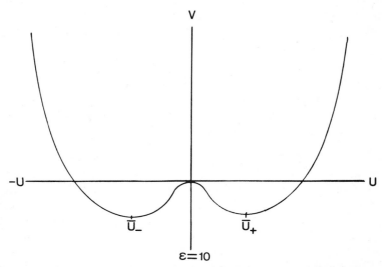

FIG. 3a. Potential function $V(U; \varepsilon, W)$ for a single homogeneous excitatory net, $\varepsilon = 10$, $w = -10$.

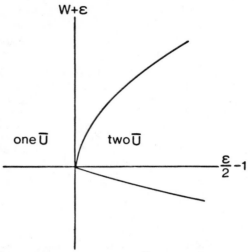

FIG. 3b. Locus of critical points in $(W + \varepsilon, \frac{\varepsilon}{2} - 1)$ plane. See text for details.

This computation suggests the use of an approximation to the potential function previously introduced. We assert that the function

$$V(U; \varepsilon, W) = \frac{1}{4} U^4 - \frac{1}{2} \left(\frac{\varepsilon}{2} - 1 \right) U^2 - (W + \varepsilon)U \quad (2.21)$$

provides the same qualitative behavior as does equation (2.17).

The associated differential equation,

$$U_T = - U^3 + \left(\frac{\varepsilon}{2} - 1 \right) U + (W + \varepsilon) \quad (2.22)$$

evidently has *qualitatively* the same critical behavior as before. The potential function defined above is known in statistical physics as the Ginzberg-Landau function, and in mathematics as the *Riemann-Hugoniot* or *Cusp catastrophe* [R. Thom 1972, 1973]. Thus (at least in the case of a single population of excitatory cells in the spatially uniform case) the generic properties of the system are well known, and follow directly from catastrophe theory. Equation (2.14) can therefore be replaced, so far as any qualitative considerations are concerned, by equation (2.22), which represents the action of a dynamical system of *corank* one (i.e., one 'state' variable U), and *codimension* two (i.e., two 'control' parameters ε and W), [Thom *op. cit.*]. Figure 4a shows the surface of equilibria $\bar{U}(\varepsilon, W)$ generated by the cusp catastrophe, and the smooth double fold in the region where there are multiple equilibria, and Figure 4b the corresponding region in the $(W + \varepsilon, \frac{\varepsilon}{2} - 1)$ plane, defined by the condition for multiple roots of $\mathrm{grad}_U V = 0$,

$$W \leqslant -\varepsilon + \left| \sqrt{\frac{4}{27}} \left(\frac{\varepsilon}{2} - 1 \right)^{3/2} \right|. \qquad (2.23)$$

It will be seen that although there is a quantitative difference between the loci depicted in Figures 3b and 4b, there is no substantial qualitative difference. Evidently a strong enough excitatory stimulus will move $\bar{U}$ from the lower level equilibrium, the 'ground-state', to the higher one, the 'excited-state', and the net will remain in such a state until quenched by a strong inhibitory stimulus. The net therefore acts as a bistable trigger-circuit, and exhibits *hysteresis* [see Andronov *et al.* 1966, E. M. Harth, T. J. Csermely, B. Beck and R. D. Lindsay 1970, and Wilson and Cowan 1972]. In the terminology of thermodynamics, the ground-state is referred to as the 'thermodynamic' branch, and the excited-state as the 'non-thermodynamic' branch, generating a so-called 'dissipative structure' [I. Prigogine and J. Nicolis 1971].

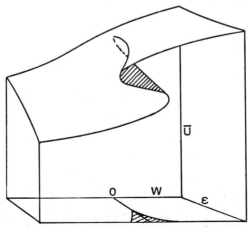

FIG. 4a. Equilibrium surface $\bar{U}(\varepsilon, W)$ generated by the cusp catastrophe.

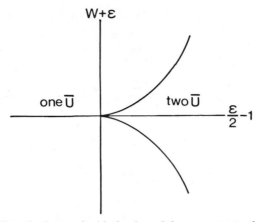

FIG. 4b. Locus of critical points of the cusp catastrophe.

The actual details of the transitions between such states show the general properties found in all switching systems. The cusp equation (2.22) can in fact be solved explicitly in the special case when $W + \varepsilon = 0$ [see H. Haken 1975]. Given $U(T) = U_0$ at $T = T_0$, the solution is

$$U(T) = \pm (1 - \varepsilon/2)^{1/2} \left[\left(\frac{1 - \varepsilon/2}{U_0^2} + 1 \right) \right.$$

$$\left. \cdot e^{2(1 - \varepsilon/2)(T - T_0)} - 1 \right]^{-1/2}, \qquad (\varepsilon < 2) \quad (2.24)$$

$$U(T) = \pm (\varepsilon/2 - 1)^{1/2} \left[1 + \left(\frac{\varepsilon/2 - 1}{U_0^2} - 1 \right) \right.$$

$$\left. \cdot e^{2(\varepsilon/2 - 1)(T - T_0)} \right]^{-1/2}, \qquad (\varepsilon > 2) \quad (2.25)$$

and at the critical point itself,

$$U(T) = \pm \left[U_0^{-2} + 2(T - T_0) \right]^{-1/2}. \quad (2.26)$$

Figure 5a shows the behavior of such solutions. Evidently $\overline{U} = 0$ is the only stable equilibrium when $\varepsilon \leqslant 2$, whereas when $\varepsilon > 2$, $\overline{U} = 0$ is unstable, and the stable equilibria are

$$\overline{U}_{\pm} = \pm \left(\frac{\varepsilon}{2} - 1 \right)^{1/2}. \tag{2.27}$$

It is of interest to note that near the critical point the solutions exhibit the well-known *critical slowing down* [Haken 1975, S. K. Ma 1976].

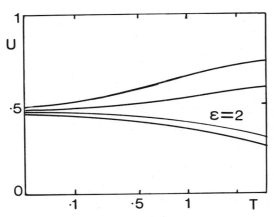

FIG. 5a. Solutions of the cusp equation.

Threshold conditions

We now consider, in more detail, the switching process itself. We assume that $W + \varepsilon = 0$ as above, and that the system is initially at $U_0 = \overline{U} = -(\frac{\varepsilon}{2} - 1)^{1/2}$, as above. Let $W = -\varepsilon = W_0$, and let there be a change $+\Delta W$ in W, so that

$$U_T = -U^3 + \left(\frac{\varepsilon}{2} - 1 \right) + \Delta W. \tag{2.28}$$

If this change persists long enough, a new equilibrium will be reached, with $\overline{U}_+$ and $\overline{U}_-$ determined by the equation,

$$U^3 - \left(\frac{\varepsilon}{2} - 1 \right)U = \Delta W. \tag{2.29}$$

Figure 5b shows the effects of such a change on the position of the curve determined by equation (2.29). It will be seen that the whole curve is simply translated downwards relative to the axis $U = 0$. Evidently if ΔW is large enough, the equilibrium $\overline{U}_-$ will be annihilated, leaving $\overline{U}_+$ as the only remaining stable equilibrium. It is easy to see that ΔW_{TH}, the *threshold* value of ΔW, equals $|\text{grad}_U \, V|$, evaluated at $U = \hat{U}$ where $\hat{U}$ is the solution of the equation $V_{UU} = 0$, i.e.,

$$\hat{U} = \pm\left[\frac{1}{3}\left(\frac{\varepsilon}{2} - 1\right)\right]^{1/2} \tag{2.30}$$

[see also equation (2.23)], whence

$$\Delta W_{\text{TH}} = 2\hat{U}_+{}^3 = 2|\hat{U}_-{}^3|. \tag{2.31}$$

A similar result obtains for the transition from $\overline{U}_+$ to $\overline{U}_-$, with a sign reversal. Thus when $\varepsilon \geq 2$, ΔW_{TH} is just large enough to move the system from $U_0 = \overline{U}_\pm$ to $U = 0$, an unstable state from which U moves to $\overline{U}_\mp$ as given by equation (2.25), and depicted in Figure 5b.

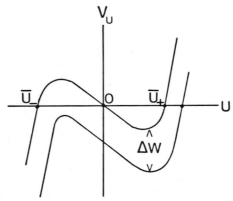

FIG. 5b. Effect of perturbation ΔW, on $\text{grad}_u V(U; \varepsilon, W)$. See text for details.

On the other hand, if ΔW is given for a very short time ΔT, then equation (2.29) results in a displacement

$$\Delta U = \left[- U_0^3 + \left(\frac{\varepsilon}{2} - 1 \right) U_0 \right] \Delta T + \Delta W\, \Delta T,$$

whence

$$U = U_0 + \Delta U$$

$$= U_0 \left[1 + \left(\frac{\varepsilon}{2} - 1 \right) \Delta T - U_0^2\, \Delta T \right] + \Delta W\, \Delta T.$$

Thus the condition for effecting a transition is simply $U \geqslant 0$, i.e.,

$$\Delta W \geqslant \Delta W_{\mathrm{TH}} = |\overline{U}_-| \cdot (\Delta T)^{-1}. \tag{2.32}$$

In general the relationship between ΔW and ΔT is given by the integral

$$\Delta T = \int_{\overline{U}_-}^0 \frac{dU}{\Delta W - U\left(U^2 - \frac{\varepsilon}{2} + 1 \right)}, \tag{2.33}$$

see D. Noble and R. B. Stein [1966], and defines ΔT as a function of ΔW. The locus of such a curve is referred to as a *strength-duration* curve [see also Wilson and Cowan 1972] and is shown in Figure 6a. The asymptotical value for $\Delta T \to \infty$, $\pm 2|\hat{U}_{\mp}^3|$ is called the *rheobase* value of the stimulus, ΔW_{RH}, and defines that stimulus strength which will *never* result in a state transition, even if $\Delta T \to \infty$. As an approximation to equation (2.33) we can use a composite of equations (2.31) and (2.32), namely

$$\Delta W_{\mathrm{TH}}\uparrow\downarrow = \pm|\overline{U}_{\mp}| \cdot (\Delta T)^{-1} \pm 2|\hat{U}_{\mp}^3|. \tag{2.34}$$

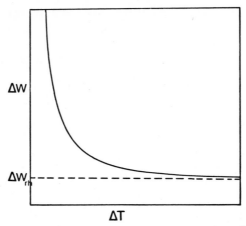

FIG. 6a. Strength-duration curve $\Delta W(\Delta T)$ for $U_T = -U^3 + (\frac{\varepsilon}{2} - 1) + \Delta W$.

The range of hysteresis

It follows that the difference $\Delta W_{\text{TH}}\uparrow - \Delta W_{\text{TH}}\downarrow$ will define the range over which such state-transitions cannot be effected, and it therefore provides a measure of the width of the 'hysteresis cycle' of the net, as shown in Figure 6b. This width is therefore given by the expression

$$h = 2|\overline{U}_-| \cdot (\Delta T)^{-1} + 4|\hat{U}_-^3|, \qquad (2.35)$$

so that $h \cdot \Delta T = 2W_{\text{TH}}\uparrow \cdot \Delta T$.

Similar results obtain in the more general case when $W_0 + \varepsilon \neq 0$. We note that $U_+ - U_- = 2\sqrt{\left(\frac{\varepsilon}{2} - 1\right)}$ independently of W, as long as W satisfies the inequality (2.23), as a consequence of the translation of $\text{grad}_U V$ relative to the U axis, for a given jump ΔW. It follows that, in general, if $W + \varepsilon \neq 0$, the threshold conditions become

$$\Delta W_{\text{TH}}\uparrow\downarrow = \pm 2|\hat{U}_\mp^3| \mp \Delta W_0 \qquad (2.36)$$

in the case of a long-lasting stimulus, where $\Delta W_0 = W + \varepsilon$. In similar fashion, for a very short stimulus duration

$$\Delta W_{\text{TH}}\!\uparrow\!\downarrow \;=\; \pm|\overline{U}_{\mp}\mp\Delta W_0|\cdot(\Delta T)^{-1}\mp\Delta W_0 \qquad (2.37)$$

leading to the strength duration curves,

$$\Delta W_{\text{TH}}\!\uparrow\!\downarrow \;=\; \pm|\overline{U}_{\mp}\mp\Delta W_0|\cdot(\Delta T)^{-1}\pm 2|\hat{U}_{\mp}^{3}|\mp\Delta W_0 \quad (2.38)$$

and to the expression for h,

$$h = \left(|\overline{U}_-\ -\Delta W_0| + |\overline{U}_+\ +\Delta W_0'|\right)\cdot(\Delta T)^{-1}+ 4|\hat{U}_-^{\ 3}|, \quad (2.39)$$

where $\Delta W_0' = -\Delta W_0$. Thus $h \to 4|\hat{U}_-^{\ 3}|$ as $\Delta T \to \infty$, i.e., the width of the hysteresis cycle, in the limit as $\Delta T \to \infty$ is independent of W, and depends only on ε.

The cusp catastrophe is clearly not only useful in providing a qualitative picture of the switching process, but it also facilitates a detailed calculation of the various threshold phenomena involved in the process.

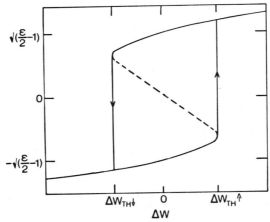

FIG. 6b. Hysteresis cycle for $U_T = -U^3 + (\frac{\varepsilon}{2} - 1) + \Delta W$.

'Nearest neighbor connectivity'

We now consider the more complex spatially dependent case, as represented by equation (2.12). The corresponding differential equation takes the form,

$$X_T - X_{Txx} - X_{xx}$$
$$= -X + 2\varepsilon[1 + e^{-(X-\theta)}]^{-1} + Y - Y_{xx}. \qquad (2.40)$$

As before we let $U = X - \theta$, but $Y - Y_{xx} - \theta = W'$, so that equation (2.40) reduces to

$$U_T - U_{Txx} - U_{xx} = -U + 2\varepsilon[1 + e^{-U}]^{-1} + W'$$
$$= -\text{grad}_U V(U; \varepsilon, W'). \qquad (2.41)$$

This reduces to the spatially constant case considered previously, when $U_{xx} = U_{Txx} = 0$.

We consider first the spatial steady state equation

$$U_{xx} = \text{grad}_U V(U; \varepsilon, W') \qquad (2.42)$$

which can be rewritten in the form

$$U(x) = \varepsilon \int_{-\infty}^{\infty} e^{-|x-y|}[1 + e^{-U(y)}]^{-1} \, dy + W', \qquad (2.43)$$

an inhomogeneous Hammerstein integral equation. We note that if $\overline{U}(x) = M$ is a equilibrium, then $U_{xx} = \text{grad}_U V = 0$, so that just as in the previous case, the extrema of the potential function V will correspond to the equilibrium solutions of equation (2.42). Once again bifurcations will occur at critical values of the parameters ε and W'. In order to gain some insight into the nature of the equilibrium solutions, we again use Liapunov's direct method, and examine the linearized equation,

$$U_{xx} + (2\varepsilon' - 1)U = 0, \qquad (2.44)$$

where $\varepsilon' = \varepsilon \dfrac{\partial}{\partial U}[1 + e^{-U}]^{-1}$, evaluated at $U = \overline{U}$, $\overline{U}$ being a solution of equation (2.15), suitably modified. When $\varepsilon' > \frac{1}{2}$, the solution is oscillatory, when $\varepsilon \leqslant \frac{1}{2}$, the solutions are either $\overline{U} = 0$ or $\overline{U} = ae^{-|x|}$. Given W' uniform, we expect $\overline{U} = 0$, but if $W' = \delta(x)$, we expect $\overline{U} = e^{-|x|}$. In what follows we assume W' uniform, but the analysis can also be carried out for more general stimuli $W'(x)$. It follows from our previous analysis of the spatially uniform case that the locus shown in Figure 3b now forms a boundary between the region I in which solutions are oscillatory and there are *no* real solutions, and the region II in which constant equilibria $\overline{U}_+ > 0$ and $\overline{U}_- < 0$ exist.

It is of interest to compare these results with those obtained via bifurcation theory [E. T. Dean and P. L. Chambré 1970], which requires simply the computation of maxima and minima of the function,

$$g(\overline{U}; \varepsilon, W') = \left[1 + e^{-\overline{U}}\right]^{-1} + W'/2\varepsilon. \qquad (2.45)$$

We first set $W' = -\varepsilon$, whence $\overline{U} = 0$. In such a case $g = 0$, $g_U = 1/4$, $g_{UU} = 0$, and $g_{UUU} = -1/4$. Under these conditions Dean and Chambré showed that for $\varepsilon < \varepsilon_c$ there are no real solutions, and for $\varepsilon > \varepsilon_c$ there are two real solutions $\overline{U}_+ > 0$, $\overline{U}_- < 0$. A similar analysis obtains when $W' \neq -\varepsilon$: thus if $0 > W' > -\varepsilon$, then $\overline{U} = M > 0$, $g > 0$, $g_U > 0$, and $g_{UU} < 0$, and again in both cases the same bifurcation behavior obtains as before. The results of course are exactly those generated by the condition $\text{grad}_U V = 0$. It follows from our previous results that the cusp catastrophe will also reproduce such results, at least qualitatively, so that equation (2.41) can be approximated by the equation

$$U_T - U_{Txx} - U_{xx} = -U^3 + \left(\frac{\varepsilon}{2} - 1\right)U + (W' + \varepsilon). \quad (2.46)$$

This equation, and the associated spatially uniform one, equation (2.39), are very closely related to the famous Bonhoeffer-van der Pol-Fitzhugh (BVP) equation used to approximate the Hodgkin-Huxley axon [R. Fitzhugh 1969, E. C. Zeeman 1972], but we remind the reader that here the equations represent the aggregate response of an entire excitatory net. As to the full time-dependent solutions of equation (2.47), it is of interest to note that were it not for the term U_{Txx}, these equations would be nonlinear reaction diffusion equations of a type currently under study by many investigators [see for example A. C. Scott 1973, N. Kopell and L. Howard 1976]. We conjecture that, qualitatively, the solutions of eqn. (2.46) and those of the equation

$$U_T - U_{xx} = -U^3 + \left(\frac{\varepsilon}{2} - 1 \right)U + (W' + \varepsilon) \qquad (2.47)$$

are similar, and that the cusp catastrophe controls the behavior of such solutions. In support of this we show in Figure 7 the results of a numerical investigation of the response of a one-dimensional excitatory net to a stimulus $W'(x, T) = (H(x + \sigma) - H(x - \sigma))\delta(T)$, where $H(x)$ is the Heaviside stop function, and $\delta(T)$ the Dirac delta function. It will be seen that the stimulus switches on the net from a low level (or negative) equilibrium to a high level one, and that the 'switch' seems to propagate, i.e., there is a travelling change of state [see for example J. Nagumo, S. Yoshizawa and S. Arimoto 1965]. Thus the solution takes the form $U(x, T) = Z(cT - \alpha x)$, where $Z(y)$ is a solution of the ordinary differential equation

$$cZ_y - a^2 c Z_{yyy} - a^2 Z_{yy} = -\text{grad}_z V, \qquad (2.48)$$

and the extrema of $V(z)$ again play a central role in determining the nature of the solutions [see J. Rinzel and J. B. Keller 1973, E. C. Zeeman 1972].

In summary, a net of coupled excitatory cells act like a bistable 'flip-flop'. If the net is coupled so that any cell is connected, on the average, to every other, then the change of state appears more or less simultaneously at every location in the net. On the other hand if there is, say, nearest neighbor connectivity, the change of state propagates with a finite velocity. It remains to perform the analysis on this case, but it seems clear that there will be a range of velocities at which the change of state will propagate in a stable fashion, as the numerical analysis seems to indicate.

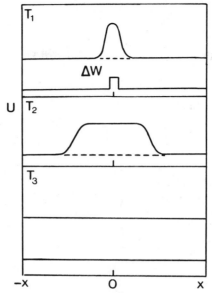

FIG. 7. Response of one-dimensional excitatory net to a brief, local stimulus.

3. NETS WITH BOTH EXCITATORY AND INHIBITORY CELLS

We now turn to a consideration of nets, the properties of which are considerably richer and more complex than those we have hitherto discussed. We first consider the simplest case of a net comprising one aggregate of excitatory cells and one of inhibitory cells, coupled together in all possible ways, as shown in Figure 8. Equation (2.12) then generalizes to

$$\frac{\partial X_1}{\partial T} = -X_1 + \varepsilon_{11} e^{-|x|/\lambda_{11}} \times \phi'[X_1]$$

$$-\varepsilon_{12} e^{-|x|/\lambda_{12}} \times \phi'[X_2] + Y_1,$$

$$\frac{\partial X_2}{\partial T} = -X_2 + \varepsilon_{21} e^{-|x|/\lambda_{21}} \times \phi'[X_1] \qquad (2.49)$$

$$-\varepsilon_{22} e^{-|x|/\lambda_{22}} \times \phi'[X_2] + Y_2,$$

where λ_{ij} is measured in units of σ, the smallest space constant in the net, and where, as before:

$$\phi'[X] = \left[1 + \exp(-(X - \theta))\right]^{-1}.$$

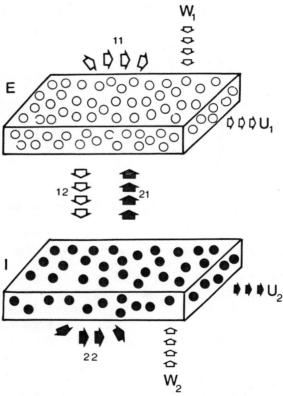

FIG. 8. Structure of a net comprising excitatory and inhibitory cell types.

All-to-all connectivity

We consider first the spatially constant case, in which there is all-to-all connectivity [see Wilson and Cowan 1972], so that equations (2.49) become, with ε_{ij} redefined to incorporate λ_{ij}:

$$\frac{\partial U_1}{\partial T} = -U_1 + 2\varepsilon_{11}\phi'[U_1] - 2\varepsilon_{12}\phi'[U_2] + W_1,$$
$$\frac{\partial U_2}{\partial T} = -U_2 + 2\varepsilon_{21}\phi'[U_1] - 2\varepsilon_{22}\phi'[U_2] + W_2, \tag{2.50}$$

where $U_1 = X_1 - \theta_1$, $U_2 = X_2 - \theta_2$, $W_1 = Y_1 - \theta_1$, and $W_2 = Y_2 - \theta_2$.

Liapunov's direct method works just as well on equation (2.50) as on equation (2.14), but the conditions are more complex. Now the equilibria are defined by the (simultaneous) equations,

$$W_1 + 2\varepsilon_{11}[1 + e^{-U_1}]^{-1} - U_1 = 2\varepsilon_{12}[1 + e^{-U_2}]^{-1},$$
$$W_2 - 2\varepsilon_{22}[1 + e^{-U_2}]^{-1} - U_2 = -2\varepsilon_{21}[1 + e^{-U_1}]^{-1}. \tag{2.51}$$

We investigate the linearized equations

$$\frac{\partial U_1}{\partial T} = -U_1 + 2\varepsilon'_{11}U_1 - 2\varepsilon'_{12}U_2,$$
$$\frac{\partial U_2}{\partial T} = -U_2 + 2\varepsilon'_{21}U_1 - 2\varepsilon'_{22}U_2, \tag{2.52}$$

where $\varepsilon'_{ij} = \varepsilon_{ij} \dfrac{\partial}{\partial U_j}[1 + e^{-U_j}]^{-1}\Big|_{U_j = \bar{U}_j}$. Following Liapunov's method, let

$$\sigma = 1 - 2\varepsilon'_{11} + 1 + 2\varepsilon'_{22}, \tag{2.53}$$

i.e., σ is minus the sum of the coefficients of U_1 in the equation for $\dot{U}_1$, and the coefficients of U_2 in the equation for $\dot{U}_2$. Similarly let

$$\Delta = (1 - 2\varepsilon'_{11})(1 + 2\varepsilon'_{22}) + 4\varepsilon'_{12}\varepsilon'_{21}, \tag{2.54}$$

the determinant of the coefficients of U_1 and U_2 in equation (2.52), and let

$$\delta = \sigma^2/4 - \Delta, \tag{2.55}$$

i.e., the discriminant of the characteristic equation $\lambda^2 + \sigma\lambda + \Delta = 0$ determining the eigenvalues of equation (2.52). Liapunov showed that there exist the following stability criteria: if (1) $\Delta > 0$, $\delta > 0$, the equilibrium is a *node*, stable if $\sigma > 0$, unstable if $\sigma < 0$; (2) $\Delta > 0$, $\delta < 0$, $\sigma \neq 0$, the equilibrium is a *focus*, stable if $\sigma > 0$ and unstable if $\sigma < 0$; and (3) $\Delta < 0$, the equilibrium is a *saddle point*.

In the above case, let

$$W_1 = \varepsilon_{12} - \varepsilon_{11}, \quad \text{and} \quad W_2 = \varepsilon_{22} - \varepsilon_{21} \qquad (2.56)$$

then $\overline{U}_1 = \overline{U}_2 = 0$, and

$$\sigma = 2 - \varepsilon_{11}/2 + \varepsilon_{22}/2, \qquad (2.57)$$

$$\Delta = (1 - \varepsilon_{11}/2)(1 + \varepsilon_{22}/2) + \varepsilon_{12}\varepsilon_{21}/4. \qquad (2.58)$$

It follows that $(\overline{U}_1, \overline{U}_2) = (0, 0)$ will be a saddle-point if $\varepsilon_{12}\varepsilon_{21}/(\varepsilon_{22} + 2) < (\varepsilon_{11} - 2)$, and either a node or a focus if $\varepsilon_{12}\varepsilon_{21}/(\varepsilon_{22} + 2) > (\varepsilon_{11} - 2)$. Since by definition $\varepsilon_{ij} > 0$, it follows that only if $\varepsilon_{11} > 2$ can $(0, 0)$ be a saddle-point, or in fact an unstable focus or node, or a stable node bounded by saddle-points. To display such possibilities we make use of the *isocline equation* [T. E. Stern, 1965],

$$M = \frac{W_2 - 2\varepsilon_{22}\left[1 + e^{-U_2}\right]^{-1} - U_2 + 2\varepsilon_{21}\left[1 + e^{-U_1}\right]^{-1}}{W_1 + 2\varepsilon_{11}\left[1 + e^{-U_1}\right]^{-1} - U_1 - 2\varepsilon_{12}\left[1 + e^{-U_2}\right]^{-1}},$$

derived from equation (2.50) in an obvious fashion. Evidently the $M = \infty$ isocline corresponds to the partial equilibrium $\dot{U}_1 = 0$, and the $M = 0$ isocline to the partial equilibrium $\dot{U}_2 = 0$. Figure 9 shows plots of these isoclines in the (U_1, U_2) plane, for particular choices of the parameters ε_{ij} and W_i. The points of intersections define the equilibria $(\overline{U}_1, \overline{U}_2)$.

It is evident from an inspection of such figures that multiple equilibria can exist only if there is a 'kink' or region of positive slope in the $M = \infty$ isocline. But the slope of this isocline is simply $-(1 - 2\varepsilon'_{11})/2\varepsilon'_{12}$, and again we obtain as a necessary

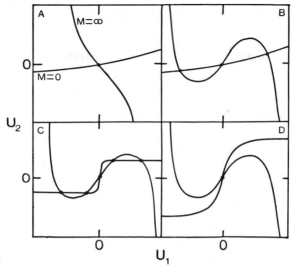

FIG. 9. Phase-plane portraits of the isoclines $M = \infty$ and $M = 0$ for an excitatory-inhibitory net. See text for details.

condition for multiple equilibria, the inequality $\varepsilon'_{11} > 1/2$, or, the sufficient condition $\varepsilon_{11} > 2$. We also note that the slope of the isocline is $2\varepsilon'_{21}/(1 + 2\varepsilon'_{22})$, and thus there are no 'kinks' in it of negative slope. So all the bifurcation behavior is controlled by the kink in the $M = \infty$ isocline. It follows that if equation (2.56) holds, then there is a single stable node (Figure 9a) if $\varepsilon_{11} < 2$, or a stable focus, for then $\Delta > 0$ and $\sigma > 0$. The node obtains if $\delta > 0$, i.e., if

$$\left(2 - \frac{\varepsilon_{11}}{2} + \frac{\varepsilon_{22}}{2}\right)^2 - (2 - \varepsilon_{11})(2 + \varepsilon_{22}) > \varepsilon_{12}\varepsilon_{21}, \quad (2.59)$$

and the focus, if the inequality is reversed. The physical meaning of this is clear. If the *positive feedback* ε_{11} is weak, but if the *negative feedback* term $\varepsilon_{12}\varepsilon_{21}$ is still smaller, relative to a measure of the positive feedback ε_{11} and ε_{22}, then the equilibrium will not have an oscillatory component, whereas if the term $\varepsilon_{12}\varepsilon_{21}$ is sufficiently strong, damped oscillatory behavior will obtain. Conversely if $\varepsilon_{11} > 2$, then multiple equilibria obtain so that $(\bar{U}_1, \bar{U}_2)$

= (0, 0) may be a saddle-point or node (Figure 9b). In the former case

$$(2 - \varepsilon_{11})(2 + \varepsilon_{22}) + \varepsilon_{12}\varepsilon_{21} < 0, \qquad (2.60)$$

the positive feedback is stronger than the negative feedback, whereas in the latter case, the above inequality is reversed, but the difference between the feedbacks is such that (2.59) obtains, and also if $\sigma < 0$, i.e.,

$$\varepsilon_{11} > \varepsilon_{22} + 4, \qquad (2.61)$$

so that the recurrent excitation is stronger than the recurrent inhibition, then the node is unstable. It follows that the phase portrait of Figure 9c, three stable nodes, obtains when $\Delta > 0$, $\delta > 0$, and $\sigma > 0$, i.e., (2.50) and (2.51) are reversed, and (2.49) holds. In this case the positive feedback dominates. Finally the phase portrait of Figure 9d, an unstable focus, obtains when $\Delta > 0$, $\delta < 0$, and $\sigma < 0$, i.e., (2.59) and (2.60) are reversed, and (2.61) holds. This case is of particular interest, because it leads to a stable *limit cycle* [Hirsch and Smale 1974]. Thus (2.61) can be redefined as a sufficient condition for the existence of such a limit cycle. This follows from Bendixon's criterion that $\sigma(U_1, U_2) = 2 - 2\varepsilon'_{11} + 2\varepsilon'_{22}$ must change sign in a single connected region R, if there is to be a closed trajectory within R [Andronov *et al.*, 1966]. This condition obtains then, when there is strong negative feedback as well as positive feedback, and the oscillatory component of the system is high. The bifurcation from a *point 'attractor'* (node or focus at (0, 0)), to a *periodic attractor* (limit cycle) occurs again at the critical value of ε_{11}, given the other appropriate conditions, and is known as the *Hopf bifurcation* [Hirsch and Smale 1974].

The behavior manifested by such nets is evidently complex. Figure 10a shows, as an example, a cross-section of the surface $\overline{U}_1(W_1, \varepsilon_{ij})$ for fixed values of the coupling parameters that result in two stable equilibria. This case corresponds exactly to that of a single excitatory net with large ε_{11}, and shows the 'fold' in the region of $W = - \varepsilon_{11} + \varepsilon_{12}$.

Similarly, Figure 10b shows a cross-section of $\overline{U}_1$ in the case when three stable equilibria are present. It is evident that three

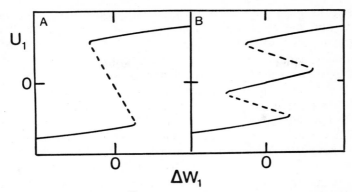

FIG. 10. Equilibrium surfaces $\overline{U}_1(W, \varepsilon)$ corresponding to the phase-plane portraits shown in Figures 9b and 9c.

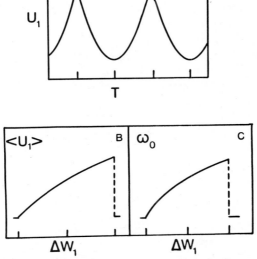

FIG. 11. Limit cycle characteristics. See text for details.

hysteresis loops can exist in such a case, one involving the ground state and the excited state, the other two involving the 'intermediate stable state'. In the limit cycle case however, there are no stationary equilibria, i.e., no stable ones, so the surface $\bar{U}_1$, does not exist. We show instead the temporal profile of the limit cycle in Figure 11a, and in Figures 11b and c, plots of respectively, $U(T)$ averaged over one period of the limit cycle, and ω_0, the limit cycle frequency, both as functions of W [see Wilson and Cowan 1972]. It will be seen that there is a *threshold* for the generation of the limit cycle, and also a *saturation-value* at which it switches off. These values of W are presumably both related to the inequality (2.23), and to the conditions on Δ, σ, and δ previously discussed.

Potential functions and catastrophes for the two component case

It is apparent that the magnitude of ε_{11} determines the various states described above. In some sense, then, the cusp catastrophe appears to underlie such phenomena. However it is difficult to see any simple way in which to demonstrate this. The problem is that equations (2.50), because of the coupling terms $\varepsilon_{12}\phi'[U_2]$ and $\varepsilon_{21}\phi'[U_1]$, no longer form a gradient system. Thus there is no longer a single potential function the extrema of which determine the equilibria, and their properties. However, equation (2.50) can be written in the following form,

$$
\begin{aligned}
\frac{\partial U_1}{\partial T} &= -\mathrm{grad}_{U_1}V_1(U_1, U_2) - \mathrm{grad}_{U_2}V_2(U_1, U_2), \\
\frac{\partial U_2}{\partial T} &= -\mathrm{grad}_{U_2}V_1(U_1, U_2) + \mathrm{grad}_{U_1}V_2(U_1, U_2),
\end{aligned}
\tag{2.62}
$$

in which there are *two* potential functions,

$$
\begin{aligned}
V_1 = \frac{U_1^2}{2} &- 2\varepsilon_{11}\left[U_1 + \ln\frac{1}{2}(1 + e^{-U_1}) \right] - W_1 U_1 \\
&+ \frac{U_2^2}{2} + 2\varepsilon_{22}\left[U_2 + \ln\frac{1}{2}(1 + e^{-U_2}) \right] - W_2 U_2,
\end{aligned}
\tag{2.63}
$$

and

$$V_2 = 2\varepsilon_{21}\left[U_1 + \ln \frac{1}{2}(1 + e^{-U_1}) \right]$$

$$+ 2\varepsilon_{12}\left[U_2 + \ln \frac{1}{2}(1 + e^{-U_2}) \right]. \tag{2.64}$$

We note that if $\varepsilon_{12} = \varepsilon_{21} = 0$, equations (2.62) do form a gradient system, whereas if $\mathrm{grad}_{U_1}V_1 = \mathrm{grad}_{U_2}V_1 = 0$, the remaining terms,

$$\frac{\partial U_1}{\partial T} = -\mathrm{grad}_{U_2}V_2, \quad \frac{\partial U_2}{\partial T} = +\mathrm{grad}_{U_1}V_2, \tag{2.65}$$

would define a nonlinear *conservative* system, with an effective 'Hamiltonian' V_2, the orbits of which in space would be closed and neutrally stable [Hirsch and Smale *op. cit.*]. In this sense the second potential function $V_2(U_1, U_2)$ defines the 'mode-mode' coupling between the excitatory and inhibitory cell types, and equation (2.62) is therefore seen to be closely related to the phenomenological kinetic equations used to represent dynamic (time-dependent) critical phenomena in statistical physics [see S. K. Ma 1976]. The reader with a knowledge of electric circuits will note that equation (2.62) represents the circuit shown in Figure 12, two non-linear resistor-capacitor circuits, coupled through a generalized *gyrator*, i.e., through a *non-reciprocal* coupler [see G. Oster, A. Perelson, and A. Katchalsky 1973]. Such a circuit acts as an asymmetric trigger circuit [Andronov *et al.*, 1966] and can be built from valves (tubes) or transistors. We remark that a large class of non-reciprocal circuits can be written in a canonical form that generalizes equation (2.62) [Cowan and Ermentrout, in preparation].

It follows that the equilibria of equation (2.62) are now determined by the equation,

$$\mathrm{grad}_{U_1}V_1 = -\mathrm{grad}_{U_2}V_2,$$
$$\mathrm{grad}_{U_2}V_1 = +\mathrm{grad}_{U_1}V_2, \tag{2.66}$$

and that at such points, $\dot{V}_1 = \dot{V}_2 = 0$. This can be proved as

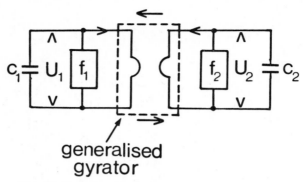

generalised
gyrator

FIG. 12. Equivalent circuit for excitatory-inhibitory nets.

follows. It can be shown from equation (2.61) that

$$\dot{V}_1 = -|\text{grad } V_1|^2 + [V_1, V_2],$$

$$\dot{V}_2 = -(V_1, V_2), \tag{2.67}$$

where $|\text{grad } V_1|^2 = \left(\dfrac{\partial V_1}{\partial U_1}\right)^2 + \left(\dfrac{\partial V_1}{\partial U_2}\right)^2$, and where

$$[V_1, V_2] = \frac{\partial V_1}{\partial U_2} \cdot \frac{\partial V_2}{\partial U_1} - \frac{\partial V_1}{\partial U_1} \cdot \frac{\partial V_2}{\partial U_2},$$

$$(V_1, V_2) = \frac{\partial V_1}{\partial U_1} \cdot \frac{\partial V_2}{\partial U_1} + \frac{\partial V_1}{\partial U_2} \cdot \frac{\partial V_2}{\partial U_2}. \tag{2.68}$$

$[V_1, V_2]$ will be recognized as the *Poisson bracket* of Hamiltonian mechanics [H. Goldstein 1952]. At any equilibrium, equation (2.66) obtains, whence equation (2.67) reduces to $\dot{V}_1 = \dot{V}_2 = 0$. The more complex conditions required to produce nodes, foci, or cycles, as expressed in equations (2.53)–(2.61), can also be rewritten in terms of V_1 and V_2. Thus

$$\sigma = -\frac{\partial}{\partial U_1} \left[-\operatorname{grad}_{U_1} V_1 - \operatorname{grad}_{U_2} V_2 \right]$$

$$-\frac{\partial}{\partial U_2} \left[-\operatorname{grad}_{U_2} V_1 + \operatorname{grad}_{U_1} V_2 \right]$$

$$= \frac{\partial^2 V_1}{\partial U_1^2} + \frac{\partial^2 V_1}{\partial U_2^2} = \nabla^2 V_1, \tag{2.69}$$

the Laplacian of V_1. Similarly

$$\Delta = \frac{\partial}{\partial U_1} \left[-\operatorname{grad}_{U_1} V_1 - \operatorname{grad}_{U_2} V_2 \right]$$

$$\cdot \frac{\partial}{\partial U_2} \left[-\operatorname{grad}_{U_2} V_1 + \operatorname{grad}_{U_1} V_2 \right]$$

$$-\frac{\partial}{\partial U_1} \left[-\operatorname{grad}_{U_2} V_1 + \operatorname{grad}_{U_1} V_2 \right]$$

$$\cdot \frac{\partial}{\partial U_2} \left[-\operatorname{grad}_{U_1} V_1 - \operatorname{grad}_{U_2} V_2 \right]$$

$$= \frac{\partial^2 V_1}{\partial U_1^2} \cdot \frac{\partial^2 V_1}{\partial U_2^2} + \frac{\partial^2 V_2}{\partial U_1^2} \cdot \frac{\partial^2 V_2}{\partial U_2^2}$$

$$= \{ V_1, V_2 \}.^* \tag{2.70}$$

We see that the condition for the existence of nodes or foci, rather than saddle-points, $\Delta > 0$, is simply

$$\{ V_1, V_2 \} > 0 \tag{2.71}$$

involving both potential functions V_1 and V_2, whereas the condition for stable nodes or foci, $\sigma > 0$, now becomes a condition on

---

*Note that in the case considered here, V_1 and V_2 are both *additive* functions of U_1 and U_2. In general $\sigma \neq \nabla^2 V_1$, and Δ is also more complex.

the Laplacian of V_1 alone,

$$\nabla^2 V_1 > 0. \tag{2.72}$$

Similarly, the condition that distinguishes between nodes and foci, $\delta = \sigma^2 - 4\Delta \gtrless 0$, also involves both V_1 and V_2. The point is that the relationship between V_1 and V_2 can be said to determine the *categorical* nature of the underlying dynamics, node, focus, saddle-point, etc., whereas the *stability* properties are solely a function of V_1. This being so, it is evident that the cusp catastrophe can again be seen to underlie the various changes of state, since only in $\mathrm{grad}_{U_1} V_1$ are there possibilities for the emergence of new equilibria. However it is ultimately the relationships between V_1 and V_2, corresponding to the various conditions on σ, and Δ, that determine the system characteristics.

In what follows we shall replace equations (2.50) by the following set:

$$\dot{U}_1 = -U_1{}^3 + \left(\frac{\varepsilon_{11}}{2} - 1\right)U_1 + (W_1 + \varepsilon_{11} - \varepsilon_{12}) - \frac{\varepsilon_{12}}{2}U_2,$$

$$\tag{2.73}$$

$$\dot{U}_2 = -U_2{}^3 + \left(\frac{\varepsilon_{22}}{2} + 1\right)U_2 + (W_2 - \varepsilon_{22} + \varepsilon_{21}) + \frac{\varepsilon_{21}}{2}U_1,$$

i.e., we approximate the potential functions of V_1 and V_2 defined in equations (2.63) and (2.64) by the functions

$$V_1 = \frac{1}{4}U_1{}^4 - \frac{1}{2}\left(\frac{\varepsilon_{11}}{2} - 1\right)U_1{}^2 - (W_1 + \varepsilon_{11} - \varepsilon_{12})U_1$$

$$+ \frac{1}{4}U_2{}^4 - \frac{1}{2}\left(\frac{\varepsilon_{22}}{2} + 1\right)U_2{}^2 - (W_2 + \varepsilon_{22} - \varepsilon_{21})U_2 \tag{2.74}$$

and

$$V_2 = \frac{1}{2}\frac{\varepsilon_{12}}{2}U_1{}^2 + \frac{1}{2}\frac{\varepsilon_{21}}{2}U_2{}^2. \tag{2.75}$$

These equations give rise to exactly the same expressions for σ and Δ, in the neighborhood of $(\overline{U}_1, \overline{U}_2) = (0, 0)$ as do equations (2.50). It remains to determine whether or not the above potential functions give the same qualitative behavior as those of equations

(2.63) and (2.64), almost everywhere in the $(\overline{U}_1, \overline{U}_2)$ plane. To do this the parametric dependence of $\overline{U}_1$ and $\overline{U}_2$ as functions of the various parameters ε_{ij}, W_{ij} is needed. It remains to compute this. However, we conjecture that the answer to this question is affirmative, and in what follows we shall use equations (2.73)–(2.75) as an appropriate representation of the excitatory-inhibitory case.

It is interesting to note that the *mode-coupling potential* we have chosen is quadratic, and therefore that the mode-coupling terms in equation (2.73) are *linear*. We note that the corresponding potential in equation (2.64) is monotonic in ε_{12} and ε_{21}, as are the functions $\text{grad}_{U_1} V_2$, $\text{grad}_{U_2} V_2$, and $\partial^2 V_2 / \partial U_1^2$, $\partial^2 V_2 / \partial U_2^2$. Moreover since ε_{12} and ε_{21} are both positive (by assumption), so are the various derivatives. Thus $V_2(U_1, U_2)$ cannot of itself give rise to any bifurcation behavior. Since the essential interactions of V_1 and V_2 all involve the second derivatives of V_1, and of V_2, any function such that $\partial^2 V_2 / \partial U_1^2 > 0$, $\partial^2 V_2 / \partial U_2^2 > 0$, which has no 'unfolding' that gives rise to changes in its extremum properties, will suffice. The function defined in equation (2.75) has a generic absolute minimum and therefore is the simplest function serving the purpose. Moreover it has the property of making the system $(U_1/\varepsilon_{12}, U_2/\varepsilon_{21})$ *anti-reciprocal* in the coupling terms, which is another way of saying that all dissipative properties are localized in $V_1(U_1, U_2)$, as required.

The potential function defined in equation (2.74) will be recognized as an unfolding of the *'double cusp'* [Thom 1973], $U_1^4/4 + U_2^4/4$. However, since the coefficient of U_2^2 in such an unfolding cannot change sign, the unfolding is really that of a single cusp, i.e., the Riemann-Hugoniot catastrophe, once again. We stress however that the global dynamics is determined, not just by this unfolding of V_1, but also by the relationships between V_1 and V_2. It remains to be seen if useful theorems can be obtained from this characterization, other than those already obtained using Liapunov's methods.

In one special case, however, an approximate reduction of the problem may be possible. Consider the case when there are no foci or limit cycles, but only nodes and (possibly) saddle-points. The appropriate condition for this, $\delta > 0$, yields the inequality:

$$\nabla^2 V_1 > 2\sqrt{\{V_1, V_2\}} \,. \tag{2.76}$$

In such a case up to 3 stable equilibria can exist. This situation is therefore described by the *butterfly* catastrophe [Thom 1973],

$$V(U) = \frac{1}{6} U^6 - \frac{1}{4} dU^4 - \frac{1}{3} cU^3 - \frac{1}{2} bU^2 - aU. \quad (2.77)$$

We conjecture that this catastrophe (in the special case when

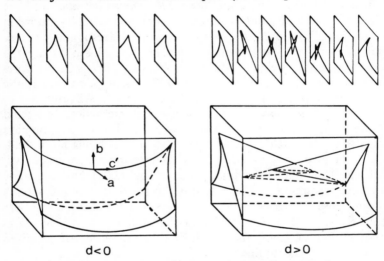

d < 0   d > 0

FIG. 13. Locus of critical points of the butterfly catastrophe.

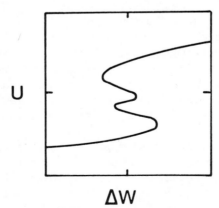

FIG. 14. Equilibrium surface $\bar{U}(\Delta W)$ generated by the butterfly catastrophe. See text for details.

$W_2 = 0$), with

$$a = W_1 + \varepsilon_{11} - \varepsilon_{12}$$
$$b = \varepsilon_{11}/2 - 1$$
$$c^2 = \sigma \tag{2.78}$$
$$d = \Delta$$

approximates the behavior of the net with both excitation and inhibition in those cases where simple bifurcations occur, but no Hopf bifurcations. The state variable U may be thought of as some appropriate combination of excitation and inhibition [J. S. Griffith 1963, J. D. Cowan 1965]. Figure 13 shows the control space of the butterfly catastrophe, in two three-dimensional sections [Thom *op. cit.*], and Figure 14, the surface $U(a, b)$ in the region $(\sigma, \Delta) > (0, 0)$, where there exist three stable equilibria, corresponding to the phase portrait shown in Figure 10c [see Wilson and Cowan 1972, and Zeeman 1972].

Nearest neighbor connectivity

We now return to equations (2.49) in which X_1 and X_2 are functions of position as well as time. As before, we let $U_i = X_i - \theta_i$, $Y_i - Y_{ixx} - \theta_i = W_i'$ and equations (2.49) become

$$\frac{\partial U_1}{\partial T} = -U_1 + \varepsilon_{11} e^{-|x|/\lambda_{11}} \times \left[1 + e^{-U_1(x)}\right]^{-1}$$

$$-\varepsilon_{12} e^{-|x|/\lambda_{12}} \times \left[1 + e^{-U_2(x)}\right]^{-1} + W_1',$$

$$\frac{\partial U_2}{\partial T} = -U_2 + \varepsilon_{21} e^{-|x|/\lambda_{21}} \times \left[1 + e^{-U_1(x)}\right]^{-1} \tag{2.79}$$

$$-\varepsilon_{22} e^{-|x|/\lambda_{22}} \times \left[1 + e^{-U_2(x)}\right]^{-1} + W_2'.$$

Once again we use Liapunov's method, and linearize about the equilibrium $(\overline{U}_1, \overline{U}_2)$, obtaining

$$\frac{\partial U_1}{\partial T} = -U_1 + \varepsilon'_{11}e^{-|x|/\lambda_{11}} \times U_1 - \varepsilon'_{12}e^{-|x|/\lambda_{12}} \times U_2,$$

$$\frac{\partial U_2}{\partial T} = -U_2 + \varepsilon'_{21}e^{-|x|/\lambda_{21}} \times U_1 - \varepsilon'_{22}e^{-|x|/\lambda_{22}} \times U_2, \tag{2.80}$$

where ε'_{ij} is defined as in the all-to-all connected case.

Fourier methods may now be used. Let

$$\hat{U}(k, T) = \int_{-\infty}^{\infty} U(x, T)e^{-ikx} \, dx. \tag{2.81}$$

Then equations (2.80) transform to

$$\frac{\partial \hat{U}_1}{\partial T} = -\hat{U}_1 + 2\varepsilon'_{11}(k)\hat{U}_1 - 2\varepsilon'_{12}(k)\hat{U}_2,$$

$$\frac{\partial \hat{U}_2}{\partial T} = -\hat{U}_2 + 2\varepsilon'_{21}(k)\hat{U}_1 - 2\varepsilon'_{22}(k)\hat{U}_2, \tag{2.82}$$

where

$$\varepsilon'_{ij}(k) = \varepsilon'_{ij}\lambda_{ij}\left[1 + \lambda^2_{ij}k^2\right]^{-1}, \tag{2.83}$$

i.e., exactly as equation (2.52) apart from the spatial frequency dependence of the coefficients $\varepsilon'_{ij}(k)$. Such equations may be thought of as giving the response of the net to sinusoidal perturbations of the form $\cos(kx)$. We note again that the infinite range of integration is an approximation to the finite range case.

As before,

$$\sigma(k) = 2 - 2\varepsilon'_{11}(k) + 2\varepsilon'_{22}(k), \tag{2.84}$$

$$\Delta(k) = \left[1 - 2\varepsilon'_{11}(k)\right]\left[1 + 2\varepsilon'_{22}(k)\right]$$

$$+ 4\varepsilon'_{12}(k)\varepsilon'_{21}(k), \tag{2.85}$$

$$\delta(k) = \sigma^2(k)/4 - \Delta(k). \tag{2.86}$$

Furthermore if the characteristic response of the net is of the form $\alpha e^{sT} \cos(kx)$, then $s = i\omega$, the 'generalized temporal frequency,' is the solution of the equation $s^2 + \sigma s + \Delta = 0$, i.e.,

$$s(k) = -\sigma(k)/2 \pm \sqrt{\delta(k)} \ . \tag{2.87}$$

The structurally stable cases [Andronov et al. 1966] are, once again, (1) $\Delta > 0$, $\delta < 0$, $\sigma \gtrless 0$; (2) $\Delta > 0$, $\delta < 0$, $\sigma \gtrless 0$; and (3) $\Delta < 0$. A detailed analysis of these cases is elementary, but tedious. A systematic study, together with the corresponding bifurcation analysis is in preparation. It suffices for the present paper to note the following qualitative properties of the solutions, obtained from a consideration of the finite range case with suitable boundary conditions: (1) There are only finitely many wave numbers k that yield eigenvalues s which are imaginary, corresponding to oscillatory temporal behavior. In fact there is a critical wave number k_c, such that all higher wave numbers yield only the trivial solution $s < 0$. (2) The number of wave numbers yielding imaginary s increases as the product $\varepsilon'_{12}\varepsilon'_{21}$ increases. Thus as $\varepsilon'_{12}\varepsilon'_{21}$ increases, the net is more likely to develop temporal oscillations, just as in the spatially constant ($k = 0$) case. (3) So far as stability is concerned, $\sigma(k) > 0$, large wave numbers are more likely to be stable. In cases where $\varepsilon'_{12}\varepsilon'_{21}$ is small, however, both small and large wave numbers are stable, and there is an intermediate range which is unstable.

Numerical integration of related equations [see Wilson and Cowan 1973] bears out the above. *Localized* states can be obtained if the space constants of the net are chosen so that

$$\sigma_{11} < \sigma_{12}, \sigma_{22} < \sigma_{21}, \tag{2.88}$$

i.e., the range of lateral interactions between cell types is greater than that of the recurrent interactions. The net effect of this choice is that any propagating activity generated in the net will be quenched by virtue of the fact that the negative feedback term $\varepsilon'_{12}\varepsilon'_{21}$ is longer-ranged than the positive feedback term $\varepsilon'_{11}\varepsilon'_{22}$. Thus if the term ε'_{11} is large, the phase portrait of Figure 9b obtains in the neighborhood of the stimulus, so that a spatially inhomogeneous stable steady state can be generated in the net, as shown in Figure 15, by brief excitatory stimuli of suprathreshold dimension. As in the spatially constant case, there is a threshold surface, defined in terms of stimulus-strength (current density), extent and duration, below which stimuli cannot switch net activity from the resting, spatially homogeneous state, to an excited, spatially inhomogeneous stable state. Thus the net functions as a spatially localized flip-flop, in effect combining computer logic with a spatially coded 'addressing' system. In addition, it is of interest that a narrow stimulus generates one localized stable pulse of activity, whereas a broad stimulus generates two stable pulses, one at each end of the stimulus. This results from the effects of lateral inhibition in the net [see F. Ratliff, 1965].

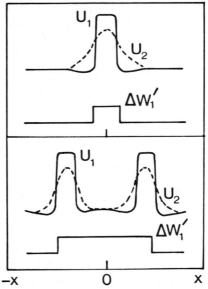

FIG. 15. Spatially inhomogeneous stable steady states of activity. See text for details.

In similar fashion, if the term $\varepsilon'_{12}\varepsilon'_{21}$ is not only long-ranged, but also sufficiently strong, then localized limit-cycles are generated in response to maintained suprathreshold stimuli, as shown in Figure 16. As in the spatially constant case, the limit cycle frequency is a function of stimulus intensity.

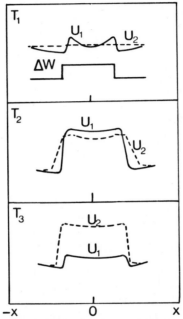

FIG. 16. Three phases in a spatially localized limit cycle in response to a maintained stimulus.

Non-localized responses can also be generated in such nets, simply by 'disinhibiting' the net, i.e., by inhibiting all the inhibitory cells in the net with a constant, low level stimulus. Under such conditions, if ε'_{11} is sufficiently strong, the situation shown in Figure 7 obtains, and the entire net will switch 'on'. However if ε'_{12} is sufficiently strong, propagating pulses will obtain, as shown in Figure 17, the trailing edge of the excitatory pulse being generated by the inhibition still present in the net.

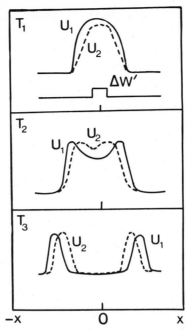

FIG. 17. Generation of travelling wave pairs in disinhibited net.

Finally, there is also another possible net response. If ε'_{11} is not sufficiently large as to produce the phase portrait of Figure 9b, but only that of Figure 9a, so that a stable node results, then it is possible to produce a long-lasting transient response, the so-called 'active' transient [Wilson and Cowan 1973], the triggering of which also defines a threshold surface. Figure 18 shows such responses. It will be seen that for given stimulus strengths, there exist both liminal durations and lengths, for the triggering of this particular response, and that the *latency* of the transient peak is related to stimulus strength.

It is evident that such nets, comprising one excitatory and one inhibitory cell type, with nearest neighbor connectivity, can generate a rich variety of responses to certain stimuli, the parameters of which may be said to be 'coded' by these nets. It is clear that even richer structure is likely to be found in two-dimensional nets, and

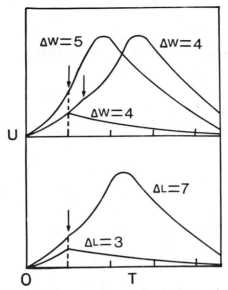

FIG. 18. Spatial and temporal summation in the transient mode.

in those nets in which the internal feedbacks generating cellular adaptation and accommodation are present, but these possibilities remain to be investigated.

4. NETS WITH MORE THAN TWO CELL TYPES

Given the complexities associated with the analysis of nets with only two cell types, it is clear that the analysis of even more complex nets is a formidable problem. However, recent work on third and fourth order chemical systems [see O. Rössler 1977] suggests that a corresponding analysis of neural nets comprising three or four cell types will prove to be of considerable interest. In this paper we merely outline the possibilities, and defer the analysis to a later paper.

We consider, once again, the pioneering works of Andronov *et al.* [1966], on the so-called "universal" electronic circuit in which both continuous and discontinuous self-oscillations are possible. Figure 19a shows such a circuit, and Figures 19b, c the corresponding three-dimensional 'flows' for, respectively, small and large values of the parameter $\mu = RC_a/rC_1$. It will be seen that in the former case the flow is continuous, whereas in the latter case it is discontinuous. Evidently as μ becomes larger, the 'kink' in the so-called *slow-manifold* increases, and discontinuous jumps, called the *fast-foliation* [see Zeeman 1972] begin to occur. As noted by Rössler [1977], if a continuous oscillator is made discontinuous by the addition of a fast foliation, and if the flow on the upper fold of the resulting slow manifold has a different orientation to that on the lower fold, then very complex *chaotic*-like trajectories can be produced. In certain cases, a new structurally stable limit set exists, termed a "strange attractor" [Ruelle and Takens 1971] the flow of which, although stable, appears to be completely chaotic.

Such observations suggest that neural switches and oscillators, when coupled together, should also display similar properties. Thus the circuit shown in Figure 20, in which an excitatory net, cell type 1, is coupled to another net comprising excitatory and

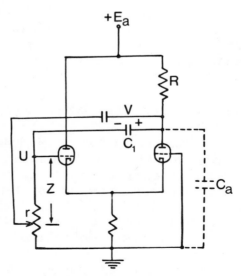

FIG. 19a. 'Universal' circuit diagram.

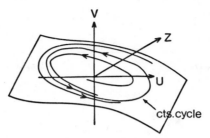

FIG. 19b. Continuous limit cycle.

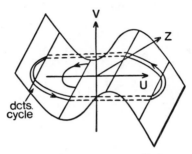

FIG. 19c. Discontinuous oscillations.

inhibitory cells, cell types 2 and 3, can function in such a fashion, given appropriate choices of the parameters. Evidently, even more complex behavior can be obtained from nets comprising 4 cell types, and so on. S. Smale [1977] has shown that related systems of 5th or higher order can exhibit almost any kind of dynamical behavior, in keeping with the intuitive notion that neural nets comprising many cell types are themselves 'universal' circuits in an extended sense. Thus a general theory of neurodynamics cannot of itself say very much, but specialized theories based upon known anatomical and physiological details can be useful. [See Feldman and Cowan 1975, Marr and Poggio 1976.] In this connection it is perhaps worth noting that such generalized malfunctions of the nervous system as the epilepsies [Jasper *et al.* 1969] may not require such details for their explication, but may be thought of as a type of neural chaos, corresponding perhaps to a quasi-strange

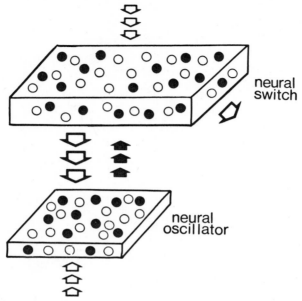

FIG. 20. Neural net that generates 'chaos'. See text for details.

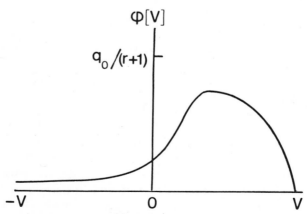

FIG. 21. Current-Voltage characteristics of a self-exciting net.

attractor solution of a multi-cell type net, in which at least one of these cell types has become intrinsically *self-exciting*, say cell type 1 of the net depicted in Figure 20, the current-voltage characteristic of which is that shown in Figure 21, rather than that of Figure 1. [See Andronov *et al.* 1966, Karzmarc 1976.]

5. CONCLUSION

We return to the problem posed in §1: the representation of events in a stable and reliable fashion, by dynamical forms arising from the cooperative and competitive activities of large numbers of coupled neurons. In §2 we showed that a net consisting only of one excitatory cell type, can be switched from one stable state to another, i.e., that such a system acts as a bistable 'flip-flop', but it cannot be locally addressed. In §3 we showed that with the introduction of inhibitory interneurons, local addressing is possible, and that local oscillators as well as multistable flip-flops can be obtained. In §4 we outlined the properties to be expected of nets consisting of three or more cell types. In general, complex chaoticlike dynamics is to be expected in nets, the structure of which is not specially evolved to limit such possibilities. It follows that in the normally functioning central nervous system, one might expect to find many inhibitory cell types (or mechanisms), the function of which is to prevent the spread of such chaotic activity, and to provide a locally addressed substrate for the representation of events via stationary or oscillating stable states, some of which may even propagate throughout the system, but in a controlled fashion. The detailed mathematical analysis of such situations remains a problem for the future. In any event the addition of a spatial or positional aspect to the mathematics of nerve nets opens up many new possibilities for the representation of events and the consequent encoding of stimulus features and forms into the structural and dynamical properties of such nets. It remains to be seen what role the dynamical aspects will play in such systems, compared with the structural aspects, but it is unlikely to be as trivial as current artificial intelligence studies suggest.

Acknowledgements. The research reported in this chapter was supported in part by the National Institutes of Health, Bethesda. We are grateful to our colleagues in the University of Chicago and the Basel Institute of Immunology, in particular Dr. H. R. Wilson, for numerous helpful comments. We should also like to acknowledge the many helpful services of D. E. Steward in the preparation of this chapter.

REFERENCES

1. A. A. Andronov, A. A. Witt, and S. E. Khaikin, *Theory of Oscillators*, Pergamon Press, London, 1966.

2. R. L. Beurle, *Philos. Trans. Roy. Soc. London Ser. B*, **240** (1956), 55–94.

3. J. D. Cowan, *Prog. in Brain Res.*, **17**, 1965.

4. ———, *Mathematics in the Life Sciences*, Amer. Math. Soc., Providence, **2** (1970), 1–57.

5. ———, *Mathematics in the Life Sciences*, Amer. Math. Soc., Providence, **6** (1974), 101–133.

6. E. T. Dean and P. L. Chambré, *J. Math. Physics*, 11, **5** (1970), 1567–1574.

7. J. C. Eccles, *The Physiology of Synapses*, Academic Press, New York, 1964.

8. ———, *Facing Reality*, Springer-Verlag, New York, 1971.

9. J. L. Feldman and J. D. Cowan, *Biol. Cybernetics*, **17** (1975), 39–51.

10. R. Fitzhugh, *Biological Engineering*, H. Schwan, ed., McGraw-Hill, New York, 1969.

11. H. Goldstein, *Classical Mechanics*, Addison-Wesley, New York, 1952.

12. J. S. Griffith, *Biophysical J.*, **3** (1963), 299–308.

13. H. Haken, *Revs. Mod. Physics*, 47, **1** (1975), 67–121.

14. E. M. Harth, T. J. Csermely, B. Beck, and R. D. Lindsay, *J. Theoret. Biol.*, **26** (1970), 93.

15. M. W. Hirsch and S. Smale, *Differential Equations, Dynamical Systems, and Linear Algebra*, Academic Press, New York, 1974.

16. A. L. Hodgkin and A. D. Huxley, *J. Physiology (Lond.)*, **117** (1952), 500–544.

17. D. H. Hubel and T. N. Wiesel, *J. Comp. Neurol.*, 158, **3** (1974a), 267–294.

18. ———, *J. Comp. Neurol.*, 158, **3** (1974b), 295–306.

19. H. Jasper, A. Ward, and J. Pope, *The Epilepsies*, Brown, Lippincott, Boston, 1969.

20. J. Karzmarc, *Biol. Cybernetics*, **22** (1976), 229–234.

21. D. Kernell, *Brain Res.*, **11** (1968), 685–687.

22. N. Kopell and L. Howard, *Mathematics in the Life Sciences*, Amer. Math. Soc., Providence, **7** (1976), 201–216.

23. J. J. Kozak, S. A. Rice, and J. D. Weeks, *Physica*, **54** (1971), 573–592.

24. J. Lund, R. Lund, A. Hendrickson, A. Bunt, and A. Fuchs, *J. Comp. Neurol.*, **164** (1976), 287–304.

25. S. K. Ma, *Modern Theory of Critical Phenomena*, Benjamin, Reading, 1976.

26. D. Marr and T. Poggio, *Science*, **194** (1976), 283–287.

27. W. S. McCulloch and W. Pitts, *Bull. Math. Biophys.*, **5** (1943), 115–137.

28. J. Nagumo, S. Yoshizawa, and S. Arimoto, *IEEE Trans. on Circuit Theory*, CT-12, (1965), 400–412.

29. D. Noble and R. B. Stein, *J. Physiology (Lond.)*, **187** (1966), 129–162.

30. G. Oster, A. Perelson, and A. Katchalsky, *Quart. Revs. Biophysics*, 6, **1** (1973), 1–134.

31. J. Pettigrew, *From Theoretical Physics to Biology*, M. Marois, ed., S. Karger, Basel, 1977.

32. I. Prigogine and G. Nicolis, *Quart. Revs. Biophysics*, **4** (1971), 107–148.

33. W. Rall, *J. Neurophysiol.*, **30** (1967), 1138.

34. F. Ratliff, *Mach Bands*, Holden-Day, London, 1965.

35. J. Rinzel and J. B. Keller, *Biophysical J.*, **13** (1973), 1313–1337.

36. O. Rössler, *Bull. Math. Biol.*, 1977, (in press).

37. D. Ruelle and F. Takens, *Comm. Math. Phys.*, **20** (1971), 167–192.

38. A. C. Scott, *Rev. Modern Phys.*, **47** (1975), 487–533.

39. F. O. Schmitt, P. Dev, and B. H. Smith, *Science*, **193** (1976), 114-120.

40. G. Shepherd, *The Synaptic Organization of the Brain*, Yale University Press, New Haven, 1974.

41. S. Smale, Preprint, 1977.

42. T. E. Stern, *Theory of Nonlinear Networks and Systems*, Addison-Wesley, Reading, 1965.

43. J. Szentagothai, *Handbook of Sensory Physiology*, VII/3B, Springer-Verlag, Berlin, 1973, 269–324.

44. R. Thom, *Stabilité Structurelle et Morphogénèse*, W. A. Benjamin, Reading, 1972.

45. ———, *Structural Stability and Morphogenesis*, W. A. Benjamin, Reading, 1973.

46. A. M. Uttley, *Proc. Roy. Soc. B*, **144** (1955), 229.

47. H. R. Wilson and J. D. Cowan, *Biophysical J.*, **12** (1972), 1–24.

48. ———, *Kybernetik*, **13** (1973), 55–80.

49. L. Wolpert, *Current Topics Develop. Biol.*, **6** (1971), 183.

50. E. C. Zeeman, *Towards a Theoretical Biology*, C. H. Waddington, ed., Edin. U. P., **4** (1972), 8–67.

SEGMENTATION, SCHEMAS, AND COOPERATIVE COMPUTATION*

Michael A. Arbib

1. THE ROLE OF A TOP-DOWN APPROACH TO BRAIN THEORY

A truly satisfying theory of any brain would place it in an evolutionary and socio-biological context. It would build upon a careful analysis of the co-evolution of the patterns of individual and social behavior which enable the animal's species to survive, and of the brain structures which enable the animal to exhibit that behavior. However, the neurophysiological and evolutionary branches of theoretical biology have been seldom conjoined. In fact, much theoretical neurophysiology can be characterized as 'bottom-up', analyzing the function of the neuron in terms of membrane properties (see the chapter by Rinzel in this volume) or the behavior of small or uniformly structured neural networks in terms of simple models of neural function (as in the chapter by Cowan). I would suggest that a *'top-down' approach to brain theory*

*The research reported in this paper was supported in part by NIH Grant No. 5 R01 NS09755-06 COM of the National Institute of Neurological and Communicative Disorder and Stroke. The paper was completed in 1975.

can provide a bridge between *'bottom-up' neural modelling* and a full-blown *evolutionary and socio-biological study*.

The problem is essentially this: near the periphery of the nervous system—a neuron or two in from the sensory receptors or the muscle fibers themselves—single-cell neurophysiology allows us to make moderately useful statements about the functions of the neural networks. Thus, in these peripheral regions, the task of the neural modeller is fairly well-defined: to refine the description of the individual neurons, and to suggest missing details about their interconnections which will allow the overall network to exhibit the posited behavior. He may also, at a more abstract level, try to analyze—as, for example, Wilson and Cowan have done—the possible modes of activity of a network of a given structure. However, as we move away from the periphery, the situation becomes less clear. A given region of the brain interacts with many other regions of the brain, and it becomes increasingly hard to state unequivocally what role it plays in subserving some overall behavior of the organism. Again, the remoteness of the region from the periphery, and the multiplicity of its connections, makes it increasingly hard to determine by the methods of single-cell neurophysiology what the 'natural' patterns of afferent stimulation of that region may be. Thus, not only are we at a loss to tell what the region does on the basis of experimentation alone, but our theoretical study of modes of response of abstract networks becomes less compelling when we must expect the input to the region to be of a highly specialized kind which is unknown to us. Finally, we may expect that central regions will often be involved in 'computational bookkeeping', so that their activity will correlate poorly with stimuli or behavior; and, in fact, their activity will be well nigh incomprehensible without an appropriate theory of computation.

It thus seems to me that we must complement the bottom-up approach to neural modelling of small or highly structured neural nets by what I may call *the 'top-down' approach to brain theory*: given some overall function of the organism which is of interest to us, we must seek to analyze how that function is achieved by the *cooperative computation* of a number of brain regions, with the corollary specification of the natural patterns of communication between regions, and thus the specification of natural inputs for each region. It should, of course, be stressed that this theory—like any science—must succeed by successive approximations. We must start with *the analysis of relatively simple functions* whose operation can be approximated by the cooperative computation of relatively few regions. As simple models of this kind succeed, we may then look at more subtle descriptions of behavior, and take further account of the modulating influence of other regions. At the same time, we can expect that developments in the evolutionary and sociobiological analysis referred to above will provide us with more sophisticated descriptions of behavior with which to confront our 'top-down' brain theory.

While about half of this chapter will be devoted to successful brain models, the other half will be more programmatic than substantive. There are few proven methods of top-down analysis of brain function. Rather, it seems to me that further success in 'top-down' brain theory will require the injection—with substantial modification!—of many of the ideas developed in the field of *artificial intelligence*. This is the field in which computer scientists and cognitive psychologists have been working (sometimes together) to design computer programs which can represent knowledge, solve problems, exhibit aspects of natural language understanding, plan, etc. Attempts to truly understand the brain's higher cognitive functions may have little success without the sort of vocabulary that workers in artificial intelligence are trying to develop; but it must be stressed that workers in artificial intelligence have paid too little attention to parallelism. In fact, of course, one of the most striking aspects of the brain is the topographical organization of its computational subsystems, and one of the major thrusts of this chapter will be to call for new concepts in a theory of *cooperative computation* which is adequate to handle this type of computational geography.

The aim of this chapter is not to give an exhaustive view of brain theory, but rather to exemplify the thesis of this introductory section. A more thorough review appears in Arbib [1975].

2. SEGMENTATION IN VISUAL PERCEPTION

For many animals, a crucial part of perception is to recognize objects—in a broad sense that includes other organisms, as well as arrays of smaller objects—to the extent that the animal is able to appropriately interact with them. Thus, a full theory must model the range of activity of the organism, and analyze the perceptual clues required to extract information appropriate to this range of interaction. However, in this section I want to focus on a much simpler observation. It is clear that most environments present an animal—let us say one with a visual system—with a complex array of stimulation which is highly unlikely to come from a spatial arrangement of objects that the animal has ever encountered before. It thus becomes important to simplify the task of recognition by breaking it up into subproblems which include breaking the scene into regions—we call this task *segmentation*—and *aggregating regions* as aspects of a single object (or as jointly constituting cues for a certain course of action, etc.). Segmentation may proceed both upward from low-level cues such as depth, color, and texture; and downward from high-level cues provided by context, or overall patterns of local features.

In the remainder of this section, we present two models of segmentation: one which is neurally based and segments on depth features; and one which evolves from work on robot vision and uses segmentation on color and texture. In Section 5, we shall study a model which uses high-level semantic information to aggregate regions into collections which constitute different portions of a single object. To place these models in methodological context: the first is an example of neural modelling; the second is in a form which suggests possible neural implementation; while the third is far indeed from a neural network implementation, and suggests, rather, the types of data structures which must be implemented in any complex perceptual system. In terms of our comments about analyzing relatively simple functions, it must be noted that each of these systems is an isolated subsystem at present. A challenge for future research is to explicate how such subsystems can work together in a system in which depth cues, color cues, and high-level semantic cues supplement each other.

2A.   SEGMENTATION ON PREWIRED FEATURES*

While each retina provides only a two-dimensional map of the visual world; the two retinae between them provide information

---

*The treatment in this section follows Arbib, Boylls, and Dev [1975, Sec. 2].

from which can be reconstructed the three-dimensional location of all unoccluded points in visual space. We indicate this in Figure 1, where the right retina cannot distinguish A, B or C ($A_R = B_R = C_R$), and where the left retina can distinguish them but cannot determine where they lie along their ray. The two retinae can actually locate them on the ray: (A_L, A_R) fixes A, (B_L, B_R) fixes B, and (C_L, C_R) fixes C. There are two problems:

The first is that our observation that the two retinae contain enough information to determine the three-dimensional location of a point in no way implies that there exists a neural mechanism to use that information. However, Barlow, Blakemore, and Pettigrew [1967], Pettigrew, Nikara, and Bishop [1968], and others find cells in visual cortex which not only respond best to a given orientation of a line stimulus, but do so with a response which is sharply tuned to the disparity of the effect of the stimulus upon the two retinae.

The second problem is that information is given about three-dimensional location of points only when the corresponding points of activity on the retinae have been correctly paired. If the only stimuli activated in Figure 1 were at the focal point and at A, then A can only be accurately located if A_L is paired with A_R—were A_L to be paired with F_R, the system would 'perceive' an 'imaginary' stimulus at W.

The main thrust of the model presented below will be to suggest how disparity-detecting neurons might be connected to restrict ambiguities resulting from false correlations between pairs of retinal stimulation. But before giving the details, let us examine some psychological data which define the overall function of the model. Normal stereograms are made by photographing a scene with two cameras, with relative position roughly that of a human's two eyes. When a human views the resultant stereogram—with each eye viewing only the photograph made by the corresponding camera—he can usually fuse the two images to see the scene in depth. Julesz [1971] has invented the ingenious technique of random-dot stereograms to show, *inter alia*, that this depth perception can arise even in the absence of the cues provided by monocular perception of familiar objects. The slide for the left eye is prepared by simply filling in, completely at random, 50% of the squares of

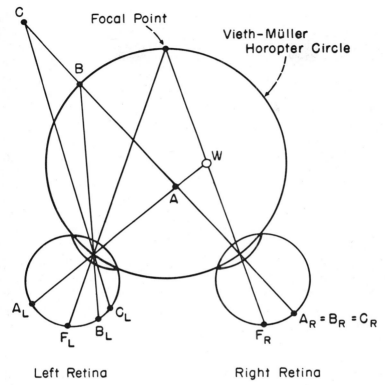

FIG. 1. The Notion of Disparity.

an array. The slide for the right eye is prepared by transforming the first slide by shifting sections of the original pattern some small distance (without changing the pattern within the section) and otherwise leaving the overall pattern unchanged, save to fill in at random squares thus left blank. With Julesz's arrays, one slide presented to each eye, subjects start by perceiving visual 'noise' but eventually come to perceive the 'noise' as played out on surfaces at differing distances in space corresponding to the differing disparities of the noise patterns which constitute them.

Note well that both stimuli of the stereogram pair are random patterns. Interesting information is only contained in the correlations between the two—the fact that substantial regions of one slide are identical, save for their location, with regions of the other slide. Then the visual system is able to detect these correlations. If the correlations involve many regions of differing disparities, the subject may take seconds to perceive so complex a stereogram— during which time the subjective reports will be of periods in which no change is perceived followed by the sudden emergence of yet another surface from the undifferentiated noise.

To clarify the ambiguity of disparity in Julesz stereograms, let us caricature the rectangular arrays by the linear arrays of Figure 2. The top line shows the 21 randomly generated 0's and 1's which constitute the 'left eye input', while the second line is the 'right eye input' obtained by displacing bits 7 through 13 two places left (so that the bit at i position goes to position $i - 2$ for $7 \leqslant i \leqslant 13$) while the bits at position 12 and 13 thus left vacant are filled in at random (in this case, the new bits equal the old bits—an event with probability $1/4$), with all other bits left unchanged. Then in the remaining 5 lines of the figure we show a disparity array, with the ith bit of the disparity of line D being a 1 if and only if the ith bit of the 'right eye input' equals the $(i + d)$th bit of the left eye input.

The disparity array of Figure 2 suggests the stripped-down caricature of visual cortex which we shall use for our model. Rather than mimic a columnar organization, we segregate our mock cortex into layers, with the initial activity of a cell in position i of layer d corresponding to the presence or absence of a match for the activity of cell i of the right 'retina' and cell $i + d$ of the left retina. (This positioning of the elements aids our conceptualization. It is not the positioning of neurons that should be subject to experimental test, but rather the relationships that we shall posit between them.) As we see in Figure 2, the initial activity in these layers not only signals the 'true' correlations (A signals the central 'surface'; B and D signal the 'background'), but we also see 'spurious signals' (the clumps of activity at C and E in addition to the scattered 1's, resulting from the probability of $1/2$ that a random pair of bits will agree) which obscure the 'true' correlations.

Let us now place this in a more general context (Dev [1975]), in which we have any set of *prewired features*—with one spatially

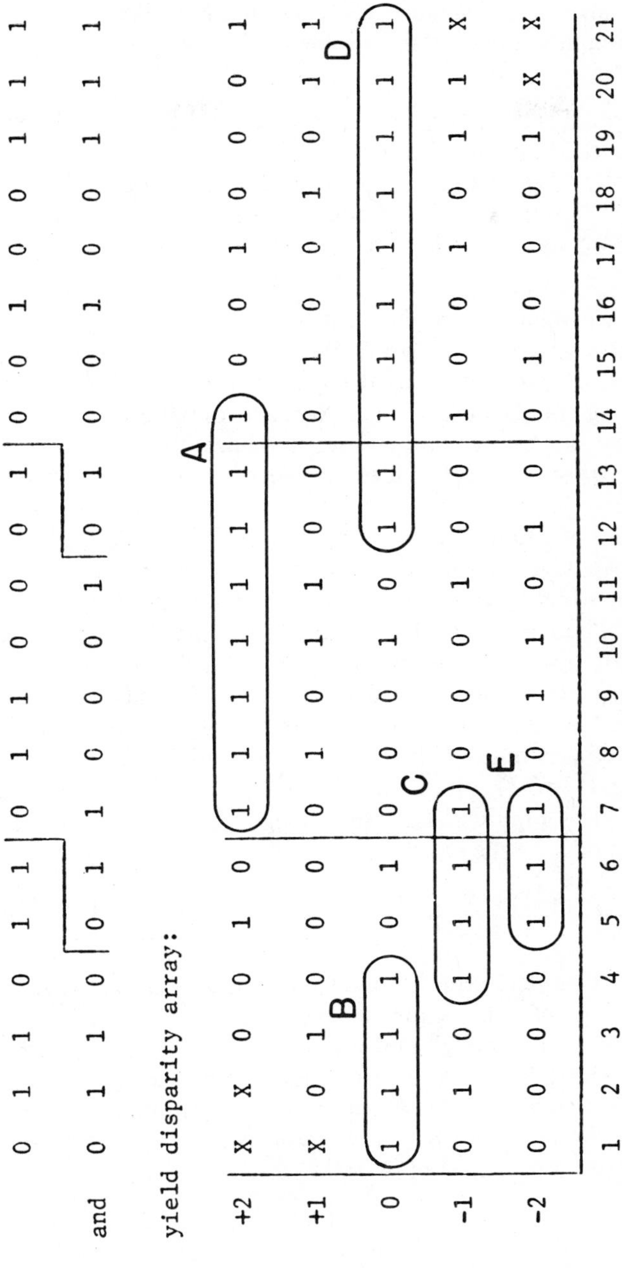

0 1 1 0 1 0 1 0 1 1 0 1 0 0 1 0 1 0 1 1 1

and   0 1 1 1 0 1 0 0 0 1 1 0 1 1 0 0 0 1 1 1 1

yield disparity array:

	1	2	3	4	5	6	7	8	9	10	11	12	13	14	15	16	17	18	19	20	21
+2	X	X	0	0	1	0	1	1	0	0	0	0	1	0	0	1	1	0	1	1	1
+1	X	X	1	1	0	0	0	1	1	1	1	0	0	1	0	0	0	0	1	1	1
0	1	1	1	0	0	1	0	1	1	1	1	1	1	1	0	1	1	0	1	0	1
-1	0	1	0	1	1	1	1	0	0	0	1	0	0	0	1	0	1	1	1	1	X
-2	0	0	0	0	1	1	1	0	1	1	0	1	0	0	1	0	0	0	1	X	X

FIG. 2. Segmentation on Disparity Cues.

coded array of detectors for each feature. We then have the following situation for the problem of segmentation of prewired features:

Conceptualization: 'Layers' of cells (they are really in 'columns'), one for each prewired feature.
Principle: Minimize the number of connected regions.
Possible Solution: Moderate local cross-excitation within layers; increasing inhibition between layers as difference in feature increases.

Can we, then, interconnect the 'layers' in such a way that clearly defined segments will form? We might imagine (but only as a crude first approximation) the resultant array of activity as then providing suitable input for a higher-level pattern-recognition device which can in some sense recognize the three-dimensional object whose visible surfaces have been so clearly represented in the brain.

While we do not have a precise functional 'region measure' which we have minimized, we do have a plausible interconnection scheme which yields qualitatively appropriate behavior of the array: The essential idea is given by the rule that there be moderate local cross-excitation within a layer; and inhibition between layers which increases as the difference in feature increases. Let then $x_{di}(t)$ represent the activity of the cell in position i of layer d at time t (where we now let activity vary continuously between 0 and 1); and let $h(j)$ and $k(j)$ be functions of the form indicated by Figure 3. Then the change of acitvity of a cell is given in our model by the equation

$$x_{di}(t+1) = \sum_{d'} \sum_{i'} h(d-d')k(i-i')x_{d'i'}(t) + x_{di}(t_0) \quad (1)$$

where it is understood that the sum 'saturates' at 0 and at 1. [**Note added in proof:** This saturation nonlinearity is crucial for correct operation of the model, and was omitted in the linear approximation of Dev [1975]. For a mathematical analysis of the nonlinearity, see S. Amari and M. A. Arbib, "Competition and Cooperation in Neural Nets," *Systems Neuroscience*, J. Metzler (ed.), Academic Press, New York, 1977, 119–165.]

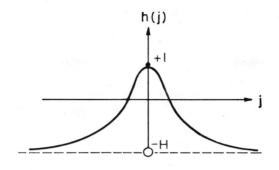

*inhibitory interaction
of layers*

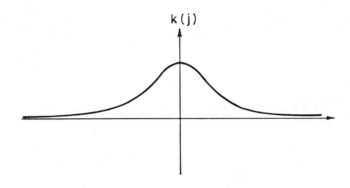

*excitatory interaction
of neighboring cells*

FIG. 3. Interaction Coefficients.

What this scheme does is allow a clump of active cells in one layer to 'gang up' on cells with scattered activity in the same region but in other layers, while at the same time recruiting moderately active cells which are nearby in their own layer. The system then tends to a condition in which the activity is clearly separated into 'regions', with each region having its own unique feature (layer of activity). In other words, such a scheme resolves feature ambiguity through suppression of scattered activity, thus permitting activity related to only one feature in any one locale. Moreover, returning to the stereopsis example, the dynamics of the model does represent the Julesz phenomenon of a noise stereogram taking some time to be perceived, with each new surface being perceived rather abruptly. This is simulated in the model by the fact that, once a sufficient number of clumps achieve high activity, the recruiting effect fills in the gaps between the clumps to form a good approximation to its final extent.

Before closing our discussion of this model, we should note that equation (1) can be rewritten in a fashion which suggests a plausible scheme of neural interconnection.

We decree that the neurons of the above array all be excitatory. We now introduce a layer of inhibitory interneurons, one for each spatial direction, the ith of which has activity at time t given by the simple equation

$$y_i(t) = \sum_d x_{di}(t).$$

Let us now pick a constant H such that $\bar{h}(j) = H + h(j) \geqslant 0$ for all j. We may then rewrite (1) in the form

$$x_{di}(t + 1) = \sum_{d'} \sum_{i'} \left(\bar{h}(d - d') - H \right) k(i - i') x_{d'i'}(t) + x_{di}(t_0)$$

$$(2)$$

so that

$$x_{di}(t + 1) = \left\{ \sum_{d'} \sum_{i'} \bar{h}(d - d') k(i - i') x_{d'i'}(t) \right\}$$

$$- \sum_{i'} 1(i - i') y_{i'}(t) + x_{di}(t_0) \qquad (3)$$

where $1(i - i') = Hk(i - i')$. Thus (3) shows that our model may be given structural expression in a form in which the x_{di} are all excitatory, with excitation being appropriately counteracted by inhibition from single layer of inhibitory interneurons.

2B. SEGMENTATION ON AD HOC FEATURES

While anyone who has used the focus control of a camera finds it plausible that a small number of different disparities can give a tolerable set of depth cues to aid other mechanisms for locating objects in the world, credulity would be strained by the suggestion that we have a prewired set of features for every color or texture which will prove of value in setting off one region of the visual world from another. In this section, then, we present a scheme which creates *ad hoc* features for segmenting visual input. It is due to Hanson, Riseman, and Nagin [1975], who give full references to related literature. The scheme is part of a preprocessor for the visual system of a robot which is to analyze outdoor scenes. In what follows, I describe a scheme of segmentation based on their model, rather than detailing their computer implementation. Some of the computations in the scheme have clear neural implementations; other parts challenge us to find neural implementations.

The system input consists of three spatially coded intensity arrays, one each for the red, green, and blue components of the visual input. The first task of the system is to extract *microtextures* —features, such as hue, which describe small 'windows' in the scene. Even in the foliage of a single tree, or in a patch of clear blue sky, the hue will change from window to window, and the system must be able to recognize the commonality amongst the variations. However, in segmenting a natural scene, macrotexture will be more important than microtexture. *Macrotexture* is a pattern of repetition of (one or more) microtextures across many local windows, and its recognition requires the analysis of structural relationships between types of microtexture. For example, the branching of a tree would have the microtexture of leaves interspersed with that of shadows in summer; while the microtextures of branches and of sky might characterize its winter appearance.

We extract microtexture first. The aim is to do this *without* using *predefined* features. The general method is as follows: pick *n* feature parameters; map each image point into the *feature space* forming an *n-dimensional histogram*. Then apply a *clustering* algorithm to segregate the points into a small number of clusters in feature space. Each cluster then forms a candidate for a microtexture of use in segmenting the original image. The point is that, for example, while tree foliage and grass may each yield a range of greens, the feature points of the two regions should form two clusters with relatively little overlap. The result of the clustering operation is suggested by the (hypothetical) example of Figure 4.

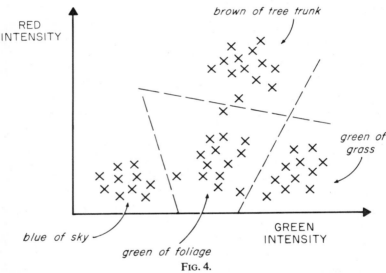

Fig. 4.

An image of a tree in a field is mapped into feature space. The four microtextures of blue of sky, green of foliage, green of grass, and brown of tree trunk define four clusters in feature space. The lines drawn to separate clusters are somewhat arbitrary, and further processing is required to settle 'demarcation disputes'.

Once the clusters have been formed, each may be assigned a distinct label. Returning to the original image, a cluster label may be associated with each point, which thus has a *tentative*, and *ad hoc*, microtexture associated with it. So far, however, no spatial information has been used to bind texture elements together. If N clusters had been formed, spatial information is used to construct an N by N adjacency matrix, in which the (i, j) element records the number of times a point labelled i is adjacent to a point labelled j (i.e., bearing the microtexture label of the jth cluster). A homogeneous region of points from a single cluster j will yield large values of the (j, j) element of the matrix. A large number of adjacencies between points of two cluster types *may* signal a cohesive region that has a repetitive mixture of cluster types—in which case it determines a microtexture. [However, note that two microtextures may be interdistributed in different ways to determine macrotextures.] Macrotextures suggested by large (i, j) entries can then be used as labels for the final pass of region-growing on the image.

The contribution from a boundary between two homogeneous regions of types i and j, respectively, would distort the (i, j) entry. To avoid this, it pays to remove large connected homogeneous regions, and then form a modified adjacency matrix without these boundary contributions. Another boundary-value problem (!) is that clustering may yield a cluster due to windows overlapping 2 regions. However, if we apply curve-following, as well as region-growing, algorithms to the cluster types, we can generate boundaries directly to supplement our region-growing process.

The resultant regions, labelled by macrotexture features, can then provide the input to a semantic labelling process of the kind we shall discuss in Section 5.

3.   COMPETITION AND COOPERATION IN NEURAL NETWORKS

Selfridge [1959] posited a character recognition system, called Pandemonium, which would behave as if there were a number of different 'demons' sampling the input. Each demon was an expert in recognizing a particular classification and would yell out the strength of its conviction. An executive demon would then decree that the input belonged to the class of whichever demon was heard yelling the loudest.

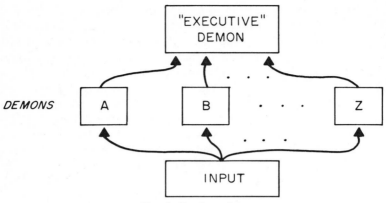

FIG. 5. Pandemonium.

On the other hand, Kilmer, McCulloch, and Blum [1969], in modelling the reticular formation, posited a system without executive control. Rather, each of an array of modules *sampled* the input and made a preliminary decision to the relative weights of different modes as being appropriate to the overall commitment of the organism. The modules were then coupled in a back-and-forth fashion so that eventually a majority of the modules would agree on the appropriate mode—at which stage the system would be committed to action. A reasonable analogy is a panel of physicians sharing symptoms and coming to a consensus about a diagnosis for a patient. (This suggests that social analogies may once again play an important role in brain theory.)

Here is the operation of the model, S-RETIC, in more detail. Each module M_i receives a sample γ_i of the input lines, with more nearby modules tending to receive more highly overlapping samples. At each time period, M_i emits as output a probability vector $p_i = (p_{i1}, \ldots, p_{im})$, where p_{ij} is the weight that M_i currently assigns to the hypothesis that the system input justifies the system committing itself to mode j of overall activity.

In addition to γ_i, M_i receives 2 input vectors p_i^H and p_i^L. Each p_{ij}^H is p_{kj} for some $k > i$ depending on i and j; similarly p_{ij}^L is a p_{kj} for some $k < i$. The k's are distributed randomly in such a way that small values of $|k - i|$ are favored.

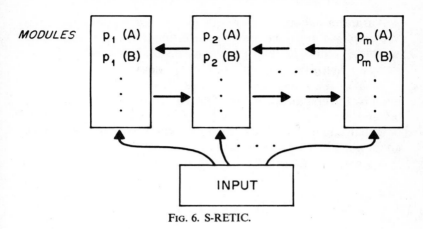

FIG. 6. S-RETIC.

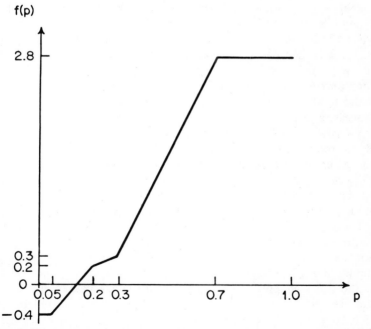

FIG. 7. This is the curve for $m = 4$; $p = 0.25$ is then the region of minimal information.

A transformation Γ_i transforms the sample γ_i into a probability vector $p_i' = \Gamma_i(\gamma_i)$—the best mode estimate on the base of the sample γ_i alone. p_i^H and p_i^L are normalized to yield probability vectors $p_i'' = N(p_i^H)$ and $p_i''' = N(p_i^L)$.

Each of the components of these 3 vectors is then operated upon by the nonlinear function f which accentuates p-values above 0.3 and diminishes those below 0.2. f implements 'redundancy of potential command'—those modules with the most information about a mode have most authority.

A secondary mode estimate is then formed by the formula

$$\bar{p}_i = \frac{C_\pi f(p_i') + C_\delta f(p_i'') + C_\alpha f(p_i''')}{C_\pi + C_\delta + C_\alpha},$$

where the C_π, C_δ and C_α are adjustable weights, such that C_π is far greater than C_δ and C_δ at times of S-RETIC change; but relax to roughly equal values thereafter. [Note the computation does not depend on the old output vector of M_i—it is as if M_i "forgets" what it has learnt unless somebody later "reminds" him of it.] The vector $\bar{p}_i$ is then passed through a transform R to form the output vector p_i. R operates by accentuating differences between small components, and shrinking between large components; adding a scalar to all components to make them positive; reversing the first process; and then normalizing: $R = N \cdot h^{-1} \cdot T \cdot h$.

We say S-RETIC converges to output mode j if more than 50% of the modules indicate the jth mode with probability > 0.5. In computer simulations, convergence always took place in less than 25 cycles; and, once converged, stayed converged for that input. Strong p_{ij} values for a given j are more likely to switch the net into mode j if the i's are close than if the i's are widely scattered.

Didday [1970], in modelling the snapping behavior of a frog confronted with two flies, posited a system of competitive interaction in the frog's tectum, which would lead in most cases to the suppression of all but one region of 'bugness' signalling, and result in the frog's snapping at one of the flies which caused the visual stimulation. In some cases, however, no region would emerge victorious from the competition. More precisely, a frog confronted with several wiggling stimuli (be they flies, or the motion of the tip of the experimenter's pencil) may exhibit one of three responses:

(a) Snap at one of the stimuli.
(b) Snap at the 'average position' of two stimuli.
(c) Snap at none of the stimuli.

On the basis of the Pitts-McCulloch model [1947] of superior colliculus, and the Braitenberg-Onesto model [1960] of cerebellum, Arbib [1972, Sec. 5.5] posited the existence of a *distributed motor controller* which responds to a pattern of high-level stimulations on its input surface by triggering a motion to the center of gravity of the spatial positions encoded by the loci of high-level inputs. [Regrettably, experimental data on type-coding vs. target-coding of actions in mammalian brain is still very indecisive.] Such a controller must be *hierarchical* since, for example, with increasing angle, a frog's head movement requires activity mainly in neck muscles, then trunk involvement, until with greatest angles hind-leg stepping is required. The above considerations lead to the scheme of Figure 8.

Our task is to give a structural description of the box M whose input-output behavior is suggested by the experiments. Namely, given a spatial input array $I(x, y)$ with well-defined peaks at $(x_1, y_1), \ldots, (x_n, y_n)$, say, M should emit an output array such that one of three cases holds:

(a) Normally, the output array will have a single large peak at the (x_j, y_j) for which $I(x_j, y_j)$ is maximal.

(b) Occasionally, the output array will have two peaks corresponding to the two largest $I(x_j, y_j)$—the motor controller will convert this into a snap at the average position.

(c) No part of the output array will reach the trigger-level for the motor controller.

In short, M resolves the redundancy of potential commands in $I(x_1, y_1), \ldots, I(x_n, y_n)$ to enable the frog to snap, normally, at one food-worthy location. Even case (b) may be valuable, as when

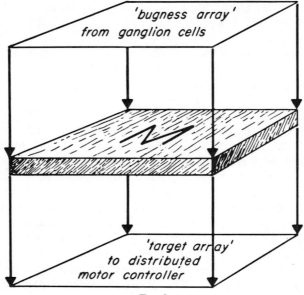

'bugness array'
from ganglion cells

'target array'
to distributed
motor controller

FIG. 8.

a frog snaps between two wiggles which are the ends of a worm. The goal was to *keep the logic distributed* rather than channeled through the serial computation of a localized executive. This, in fact, Didday achieved using two layers of cells, whose names suggest their relation to cell types actually observed by Lettvin *et al.* [1960]:

  (i) The *sameness* cells: sum the total 'bugness' activity outside their own region.
  (ii) The *newness* cells: signal change in 'bugness' in an area.

  More specifically, let

$f(x, y, t)$ = the 'bugness' evaluation made by the tectum on the basis of ganglion cell activity; at position x, y; at time t.

$m(x, y, t)$ = a 'masked' evaluation of bugness—the output of M at position x, y; at time t.

$s(x, y, t)$ = the activity of the sameness cell at position x, y; at time t.

$n(x, y, t)$ = the activity of the newness cell at position x, y; at time t.

These activities are then related as follows:

$$s(x, y, t + 1) = \left[\sum_{x', y' \notin B_{x,y}} m(x', y', t) \right] \Big/ \left[1 + \sum_{x', y' \in B_{x,y}} m(x', y', t) \right]$$

where $B_{x,y}$ is a small region around (x, y)—i.e., $s(x, y, t + 1)$ is large to the extent that most m activity is outside $B_{x,y}$.

$$m(x, y, t + 1) = n(x, y, t + 1)$$
$$+ \left\{ m(x, y, t) / \left[1 + h\big(s(x, y, t + 1)\big) \right] \right\}$$

$$\text{where } h(s) = \begin{cases} 0 & \text{if} \quad s < .2 \\ s & \text{if} \quad .2 < s \leqslant 1.6 \\ 2s - 1.6 & \text{if} \quad s \geqslant 1.6 \end{cases}$$

so that m remains unchanged where $B_{x,y}$ contains a large proportion of the 'masked bugness'; and is drastically reduced where $B_{x,y}$ contains very little,

while

$n(x, y, t + 1)$ = increase in $f(x, y, t)$ over $f(x, y, t - 1)$ if this is positive unless the cell is habituated—a complication we ignore here.

Thus in the nonhabituated frog a new pattern is entered into the masked cell layer; thereafter (for constant input) the distributed computation iterated through the sameness cells suppresses all but the highest peak of input activity as represented in the masked cell layer. The drawback with this model is that the initial values entered by the newness cells are larger than the final values to which the $m(x, y, t)$ converge, so that one must assume that newness activity inhibits any motor effect of $m(x, y, t)$ until after a convergence period. Normally the converged pattern has only one sharp peak—whether or not it is above threshold determines whether case (a) or (c) obtains. In some cases, two nearby peaks in f coalesce in m, and we have case (b).

In attempting to place these studies in perspective, Montalvo [1975] observed that we could analyze the Dev, Didday and S-RETIC models within a common framework, with the computational subsystems arrayed along two dimensions, one of competition and one of cooperation, as in Figure 9. For further analysis, see the paper by Amari and Arbib [1977], referred to above.

DIDDAY

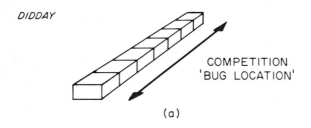

COMPETITION
'BUG LOCATION'

(a)

KILMER–McCULLOCH

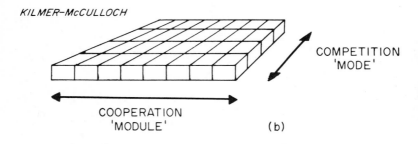

COMPETITION
'MODE'

COOPERATION
'MODULE'

(b)

DEV

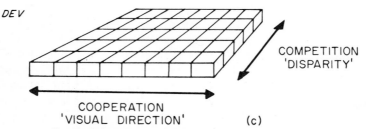

COMPETITION
'DISPARITY'

COOPERATION
'VISUAL DIRECTION'

(c)

FIG. 9. Competition and Cooperation in three neural nets.

The theme of competition and cooperation has thus emerged in three completely separate neural network models. As we shall see in the next two sections, it also plays a role when we look at the way in which an internal model of the world would operate.

4. REPRESENTING THE WORLD AS AN ARRAY OF ACTIVE, TUNEABLE SCHEMAS

We have stressed that an organism's sensory stimuli should be analyzed as an array of familiar 'objects'—whether the object was a single object, such as a tree, or a composite object such as a row-of-trees. Moreover, the representation of this array should be easily updateable. This 'array representation' is reminiscent of the making of movie cartoons, where each frame is obtained by photographing the contents of a 'slide-box'—in which are appropriately positioned slides. A background slide may remain the same for many successive frames; a middle-ground slide (representing, say, a tree) may require no redrawing, though it may require repositioning relative to the background; whereas a foreground slide (representing, say, a person) may remain the same in gross features from frame to frame yet require constant redrawing of details (to represent, for example, the movements of mouth and limbs).

The *basic slide-box metaphor* (Arbib [1970; 1972, p. 92]), then, requires the representation of input patterns to be given as an array of *slides*—appropriately located and 'tuned'—chosen from a *slide-file*, a collection of standard slides, with the array 'covering', in some suitable sense, the input. However, rather than thinking of slides in visual terms we must regard them as cueable by various modalities—the slides are to cover the 'sensible environment'—so that a cat-slide may be activated as much to cover a meow as to cover the sight of a cat. More importantly, though, from an *action-oriented* point of view, the activation of a slide is now to be seen as giving access to programs for guiding possible interaction of the organism (be it human, animal or robot) with the 'object' which the slide represents.

The important point is that we wish to analyze scenes which extend over time, and in the slide-box representation a frame can often be obtained from its predecessor by simply relocating the 'active' slides (i.e., those currently in the *slide-box*) and updating some of their parameters—rather than resegmenting the input and assigning tags to these regions. Of course, when some region is no longer tolerably covered by the active slides, some appropriate retrieval mechanism must replace one or more slides by some new slide from the slide-file, and tune it appropriately.

But this language smacks too much of the metaphor. Let us replace slide by the notion of a *schema*—an array of programs to analyze a segment of the input to determine a possible course of action. As such, a schema must be relocatable, tuneable, and linkable with other schemas. Thus, we have the following components of a schema:

(i) *Input-Matching Routines*: A routine which succeeds to the extent that it covers a spatial region of multi-modal input (so that a cat schema can cover a furry region which emits meows, but not one that emits barks). This can be biased by non-sensory context inputs. The input-matching routines may include calls for confirming information (as in the eye-movement calls of Didday and Arbib [1975]). The level of success of these input-matching routines may be regarded as an 'activation' level of the schema. The activation level of the schema increases as the location and other parameters of the schema are adjusted to better fit the covered region. However, to complicate the story, the *resolution level* (i.e., the precision of this parameter match) required for activation to saturate may well depend on goal settings, or other non-sensory input to the schema—consider the different level of precision with which we 'perceive' a tree when we intend to walk around it, as against when we intend to climb it (so that the placement and estimated strength of the branches, rather than the general area they occupy becomes important).

(ii) *Action Routines*: To the extent that the success of the input-matching routines raises the activation level of a schema to that extent do certain actions become appropriate for the organism. Programs for some of these actions then form part of the schema. A crucial integrative property of schemas is that *increasing accuracy of parameter adjustment by the input-matching routines automatically adjusts parameters in the action routines in such a way that the action becomes more appropriate for the current environment and goal structure* (if the schema has been properly 'evolved'). A classic example is that the cat turning its head to gaze at a mouse is automatically tuning its motor system for the pounce.

(iii) *Competition and Cooperation Routines*: To date, we have talked of a schema as acting in isolation, attempting to raise its activation level by proper matching of input. As we shall discuss in the next section, the operation of competition and cooperation routines helps determine which population of parameter-adjusted highly active schemas will constitute the current model of the environment. Of course, there still remains the problem of determining which of the action routines of these schemas are to operate—and another network of competition and cooperation routines will be involved in determining a compatible set of actions. This expository separation—scene analysis first, planning second—is misleading. At any time, the organism is engaged in some activity, even if that be resting. Thus it is not so much a matter of choosing a course of action as it is of determining whether the time has come to change the course of action being pursued. The completion of an action may remove it from the competition. More interestingly, the execution of an action may provide new sensory input which de-activates the slide (or drastically changes the parameter setting) which enabled the action—as when we bite into what appears to be a piece of fruit only to discover that it is made of wax.

Notice that all these routines provide the *semantics* of a schema —what an object *means* to us comprises our knowledge of what we can do with the object and what relations it has with the object, to the extent that our input-matching routines can capture the effects of these actions and relationships. In any case, we have come a long way from the original notion of a slide as being simply a colored transparency that approximates a region of the visual field. Now that we have our new general notion of a *schema* we shall henceforth reserve the word for its technical sense of an *(input-matching; action; cooperation and competition) set of routines*, with the crucial relationship between the parameters of the input-matching and action routines.

We must now look more carefully at the structure of parameter sets. Minsky [1975] notes that a person cannot visualize a cube in any perspective with real accuracy. However, I do not accept his suggestion that we can internally represent a small population of *precise* parameter settings, i.e., a precise view of cubes oriented at $-45°$, $-30°$, $-15°$, $0°$, $15°$, and $30°$ about a vertical axis. Rather, it seems to me better to explicitly regard this as an example of crude parameter setting—the routine is sloppy enough that it will, for example, accept any cube which is *roughly* head on, say from $-10°$ to $10°$ as satisfying a given parameter setting in the input-matching routine. More interestingly, we may imagine *levels of precision*, so that one range may be more precise than another. We thus require that the parameter sets in both input-matching and action routines be *partially ordered sets* (*posets*, for short), where the relation $x \sqsubseteq x'$ is to be interpreted as *x' is a refinement of x*, or as *x approximates x'*. For example, if Y is a set of reachings, we might have

reaching to the left

$\sqsubseteq$ reaching about 60° to the left

$\sqsubseteq$ reaching 62° to the left

while it is not true that

reaching 61° to the left

$\sqsubseteq$ reaching 62° to the left.

We also assume that each set has a *minimal element* 0

$0 \sqsubseteq y$ for all $y \in Y$

corresponding to 'no specification at all'—so that, in the reaching set, 0 just means 'reaching', without any specification as to the direction (and so is not to be confused with the very precise 'reaching 0° to the left', i.e., straight ahead).

For now, we shall assume that *all schemas may continually monitor their input pathways* (though different schemas have different input sets). In other words, the slide-file of the original metaphor becomes the total population of (relatively high-level) schemas of the present model; the slide-box of the original metaphor becomes the subpopulation of highly activated schemas of the present model. As in both Pandemonium and S-RETIC (recall Section 3), we let each schema (i.e., mode-element) continually receive input. However—unlike both Pandemonium and the original slide-box metaphor—we shall for now try to do without a central executive overseeing the activation of schemas and instead —in the spirit of Dev, Didday and S-RETIC—explore what can be achieved by the schemas themselves by virtue of their cooperation and competition routines. My methodological point is that it is not helpful to make *a priori* assumptions (whether to fit our preconceptions about neural net structure or about the utility of LISP programming) when setting up a framework of this generality. When we actually look at restricted systems which must be implemented in a brain or on a computer, then we can be more specific about the sets of executive and book-keeping routines that seem necessary to augment the routines built into the schemas themselves.

With this, the time has come for a formal notion of *scene* to replace the "contents of the slide-box" of the old metaphor: A *scene* (A, e) is a set A of schemas together with a function

$$e: T \to \prod_{a \in A} P_a$$

where T is a time interval, P_a is the poset of parameter-settings of schema a, and for each $t \in T$,

$$e(t) = \{e(t)(a) | a \in A\}$$

is such that $e(t)(a) \in P_a$ is the parameter-setting of schema a at time t during scene (A, e).

Without being precise, it is part of the concept that the set A is relatively small—the scene (which is a potential new 'superscheme' in our unpublished learning theory for schemas) is the internal representation of a relatively homogeneous episode (i.e., one in which there are no 'dramatic' changes in the set of schemas or their parameter settings). This does not require all schemas in A to be active throughout the scene—$e(t)(a) = 0$ is permissible—but it does require that the variation of $e(t)(a)$ over time not change the structure of the situation too drastically (as would be the case if you were suddenly to perceive that a lion had entered your room).

Note that we have changed from the usual definition of scene—a two-dimensional visual input—to a definition that is *system-dependent* (different observers may activate different sets of slides in response to a given visual input) and *extends over time*. This is consistent with our general theme of *action-oriented perception*—a scene is to be a meaningful episode in the life of an organism interacting with a dynamic environment.

It is beyond the scope of this article to relate schemas to other approaches in the literature—but a few comments are in order. The problem of representation of knowledge has long been of great interest to psychologists. A classic study is Bartlett's [1932] analysis of "Remembering"; and we have gained much from Piaget's studies of schemata, especially his notions of assimilation and accommodation—see Furth [1969] for a review. These ideas have recently been born anew in approaches to artificial intelligence. Perhaps the best-known study is that of Winograd [1972, 1973], which works with a restricted 'blocks world'. Several other approaches to this general area are presented by Schank and Colby [1973]. Minsky [1975] has recently advanced his concept of 'frames' as a unification of these studies. At the level of vision and manipulation, these ideas seem bettered by our notion of a schema. However, the idea of a schema is not yet well tuned to the problems of linguistic and social interaction. Intriguingly, the notion of frame analysis proves not to be peculiar to artificial intelligence. Erving Goffman's [1974] "Frame Analysis: An Essay on the Organization of Experience" is a text in social psychology, and many of his examples are surprisingly reminiscent of Minsky's. A contribution to this area which seems to lie intermediate between the straight A. I. approach and Goffman's approach is that of Bruce and Schmidt [1974], which can be viewed as a sequel to Searle's [1965] study of 'speech acts'. Schank's work—and the related work of Abelson [1973]—are also interesting in this regard. **[Note added in proof:** U. Neisser, *Cognition and Reality*, W. H. Freeman, San Francisco, 1976, bases his approach to cognitive psychology on a notion of schema which approximates our notion of a scene as a collage of activated schemas. For a different view of schemas, see various articles in G. E. Stelnach (ed.), *Motor Control: Issues and Trends*, Academic Press, New York, 1976.]

5. COMPETITION AND COOPERATION BETWEEN SCHEMAS

Imagine that a segmentation program has divided a scene into regions such as those shown in Figure 10.

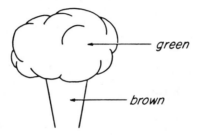

FIG. 10. What is it?

With only this much information available, two quite different pairs of schemas may be activated to cover this input: in the first interpretation the schemas would represent green ice-cream and a brown ice-cream cone; in the second interpretation, the schemas would represent the foliage and trunk of a tree. There would be competition between the pairs, and cooperation between the schemas within each pair. Thus the system of interactions shown in Figure 10 would have two large attractors corresponding to the two natural interpretations, and very small attractors for the "unreal" pairings—though these could be forced by a trick photograph or a Magritte painting.

In case of blurred images, the input may start the system close to the boundary between the two attractors. Convergence may be slow, and even incorrect. Context can provide a mechanism to speed, and correct, convergence. Thus the 2 contexts in Figure 11a correspond to the 2 extra schemas in Figure 11b, and in each case we expect rapid convergence to the "right" interpretation. Thus an initial configuration in which the schemas for foliage, trunk,

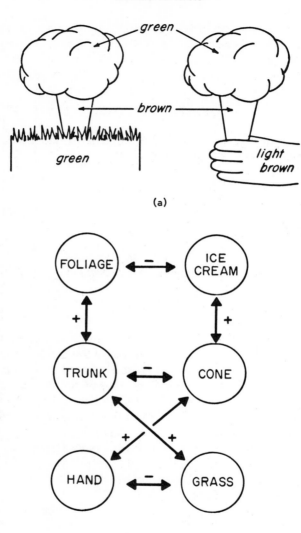

(a)

(b)

FIG. 11. The effect of context.

ice-cream and cone have comparable activity will rapidly converge to a state of high activity in foliage and trunk schemas and low activity in ice-cream and cone schemas if the grass schema is given a higher activity level than the hand schemas and vice versa. Incidentally, we may note that as well as slides for objects, we may also have more abstract schemas such as one for winter. Now at the change of seasons, the first fall of snow may be the signal for winter—so that we must posit the activity level of the snow-schema as providing excitatory input to the winter-schema. However, in the normal course of events, the organism *knows* that it is winter, and can use this *contextual information* (Figure 12) to favor the hypothesis that a white expanse is snow rather than burnished sand, say, or moonlit water. It is this type of reciprocal activation (whether we regard it as an additional input, or as the action of a cooperation routine) that gives the system of schemas its *heterarchical* character. [Strictly defined, a 'heterarchy' is a system of rule by alien leaders. But in AI, stimulated by McCulloch [1949], it now denotes a structure in which a subsystem A may dominate a subsystem B at one time, and yet be dominated by B at some other time.] This notion of schema competition and cooperation may be seen to apply to the Dev model (Section 3) which we can recast in terms of 7 arrays of schemas (Figure 13).

In this simple model, the schemas in layers 1L and 1R are simply ganglion cells of the retina, whose firing level is high to the extent that some feature is present in its visual field. The activity of a schema in layer Dk represents the presence of a feature with a given 3-dimensional location (as coded by the disparity between the schemas in 1L and 1R which activate it). But much of the activity here would be spurious if these schemas were driven only by the schemas of 1L and 1R—and so Dev postulates competition and cooperation routines (schemas in different D-layers compete; nearby schemas in a given D-layer cooperate) which yield segmentation of the visual field into relatively few regions in each of which the active D-schemas have the same feature.

Burt [1975] has modified the Dev model to support moving regions in response to moving inputs. In a more elaborate model one would then posit that—at this level—the visual field is not simply segmented into regions of relatively homogeneous schemas with high activation, but that the cooperation routines have

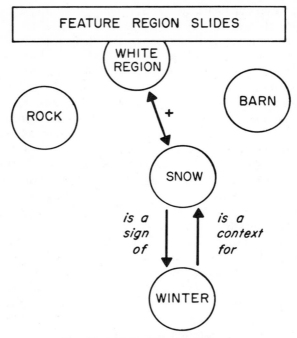

FIG. 12. A Heterarchical Relationship.

activated pointers to the activated schemas of the same region. Thus 'higher-level' schemas may determine that they are dealing with a region rather than an isolated feature.

Let us now leave this general discussion, and try to put competition and cooperation between schemas on a formal level. In one approach, due to Waltz [1975] and given a parallel algorithm by Rosenfeld, Hummel and Zucker [1976], the activation levels are 0 or 1, and one looks for a consistent labelling of the regions. In the second approach, also due to Rosenfeld *et al.*, one imposes a nonlinear relaxation scheme on assignments of probabilities to the different labelling hypotheses for each region. The latter approach, as we shall see, is more satisfactory. Even more interesting, however, is its strong formal resemblance to the S-RETIC scheme of section 3. Thus our study of competition and cooperation between

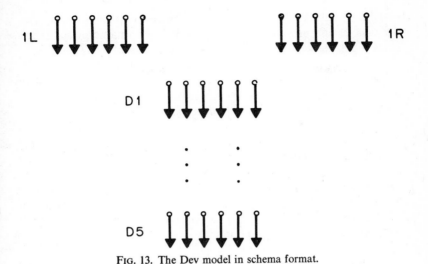

Fig. 13. The Dev model in schema format.

schemas is brought firmly within our scheme of competition and cooperation in neural networks.

The Binary Model: We are given a set $A = \{a_1, \ldots, a_n\}$ of objects to be labelled, and a set $\Lambda = \{\lambda_1, \ldots, \lambda_m\}$ of labels. Let Λ_i be the set of those labels in Λ compatible with a_i; and let $\Lambda_{ij} \subset \Lambda_i \times \Lambda_j$ be the set of pairs of labels (λ, λ') such that λ may occur on a_i when λ' occurs on a_j. We set $\Lambda_{ii} = \Lambda_i \times \Lambda_i$.

[In our motivating example, we have 3 regions, with $\Lambda_1 = \{$foliage, ice-cream$\}$, $\Lambda_2 = \{$trunk, cone$\}$, $\Lambda_3 = \{$grass, hand$\}$, while $\Lambda_{12} = \{$(foliage, trunk), (ice-cream, cone)$\}$, etc.]

A **labelling** $\mathbf{L} = (L_1, \ldots, L_n)$ assigns a set $L_1 \subset \Lambda_1$ of labels to each a_i. The labellings form a lattice under componentwise set-inclusion.

We say that a labelling is **consistent** if

$$(\{\lambda\} \times L_j) \cap \Lambda_{ij} \neq \varnothing \text{ for all } \lambda \in L_i. \tag{1}$$

This is a *local* consistency condition. We call a labelling $\mathbf{L}$ **unambiguous** if it is consistent and each $|L_i| = 1$. Note that a consistent labelling might not contain an unambiguous labelling. However, the task of the present model is to find such unambiguous labellings when they exist. We first note the obvious:

PROPOSITION: *The empty labelling is consistent. The union of any set of consistent labellings is again consistent. There is thus a greatest consistent labelling* $\hat{L}$ *(which may be null).*

Returning to the criterion, (1), for consistency let us say a label λ in set L_i is **isolated** in **L** if $(\{\lambda\} \times L_j) \cap \Lambda_{ij} = \varnothing$ for *any j*. It is clear that such a λ cannot be part of a consistent sublabelling of **L**. This suggests that we operate on **L** with Δ where Δ**L** is obtained from L by discarding from each L_i all labels which are isolated in **L**. It is clear that Δ**L** = **L** iff **L** is consistent. More interestingly, a standard fixed-point argument shows:

PROPOSITION: *For any* **L***, let* $\mathbf{L}^{(\infty)}$ *be the greatest consistent labelling contained in* **L***. Then* $\mathbf{L}^{(\infty)}$ *is the greatest fixed point of* Δ *contained in* **L***, and, by the finiteness of* Λ*, we have that there exists an integer k such that*

$$\mathbf{L}^{(\infty)} = \Delta^k \mathbf{L}.$$

To find an *unambiguous* labelling contained in **L**, we build a tree (in the style of Waltz [1975]) of labellings, with $\mathbf{L}^{(\infty)}$ as the root. Then, having obtained a node with labelling $\overline{L}$, we construct its descendants by making all possible choices as follows:
Pick an *i* such that $\overline{L}_i$ is not a singleton. Pick $\lambda \in \overline{L}_i$. Set

$$\mathbf{L}' = \left(\overline{L}_i, \ldots, \overline{L}_{i-1}, \{\lambda\}, \overline{L}_{i+1}, \ldots, \overline{L}_n \right).$$

If $\mathbf{L}'^{(\infty)} \neq (\varnothing, \ldots, \varnothing)$, add it to the tree as a descendant of $\overline{L}$.
It is clear that all unambiguous labellings (if there are any) contained in **L** will occur as (some of) the terminal nodes of the tree grown in this way.
The main problem with this algorithm is that if any region has an empty label set L_i in **L**, then Δ**L** will be the empty labelling. Clearly, a better algorithm would use the 'consensus' of other

regions to *add* labels to L_i—as when we suddenly perceive the nature of an object purely on the basis of its context. The next algorithm, then, allows continuously varying weights to be assigned to each label for each region, and uses an operation which can *increase*, as well as decrease, those weights. As remarked before, the scheme is strongly reminiscent of S-RETIC, though here convergence is towards a labelling, rather than towards consensus on a single mode.

The Nonlinear Probabilistic Model: We again have a set $A = \{a_1, \ldots, a_n\}$ of regions, and a set $\Lambda = \{\lambda_1, \ldots, \lambda_m\}$ of labels. However, a **labelling** $\mathbf{p} = (p_1, \ldots, p_n)$ is now a sequence of probability vectors $p_i : \Lambda \to [0, 1]$, with $p_i(\lambda)$ being the weight assigned by $\mathbf{p}$ to the hypothesis that λ is the correct label for a_i.

We wish to design an operator F which—in the style suggested by our discussion of Figure 11—will on iterated application move $\mathbf{p}$ towards a 'correct' labelling. The key idea is that the probability $p_i(\lambda)$ of a given label for a_i should be increased (respectively, decreased) by F if other objects that have high probability labels are highly compatible (respectively, incompatible) with λ at a_i.

Thus, in the present model, the Λ_{ij} are replaced by *compatibility functions*

$$r_{ij} : \Lambda \times \Lambda \to [-1, 1]$$

which function like correlations: if λ' on a_j frequently co-occurs with λ on a_i, then $r_{ij}(\lambda, \lambda')$ is positive; if they rarely co-occur, $r_{ij}(\lambda, \lambda')$ is negative; and if their occurrences are independent, $r_{ij}(\lambda, \lambda') = 0$.

[It is clear that the memory structures required to produce the compatibility functions may be quite elaborate. Returning to our example of Figure 11, the system would have to use the observation that regions 1 and 2 were contiguous, with region 1 above region 2, to obtain estimates like

r_{12} (foliage, trunk) $= 0.7$,
r_{12} (foliage, cone) $= -0.8$,
r_{12} (ice-cream, cone) $= 0.9$, etc.

Incidentally, the very arbitrariness of these three numbers makes it clear that the F we are constructing must be *structurally stable*—small changes in the r_{ij}'s must rarely perturb convergence. Unfortunately, we do not yet have rigorous proofs of convergence —though computer simulations are encouraging—let alone structural stability.]

To satisfy the 'key idea', we define the 'change operator' $\mathcal{K}$ by

$$(\mathcal{K}\,\mathbf{p})_i(\lambda) = \sum_j \sum_{\lambda'} r_{ij}(\lambda, \lambda') p_j(\lambda').$$

Each $\sum_{\lambda'} r_{ij}(\lambda, \lambda') p_j(\lambda')$ expresses the 'consensus' of the labelling of a_j by $\mathbf{p}$ as to the direction in which $p_i(\lambda)$ should shift.

With this definition of $\mathcal{K}$, one possible choice for F is then

$$F\mathbf{p} = \mathcal{R}[\mathbf{p} + \mathcal{K}\mathbf{p}].$$

where the normalization operator $\mathcal{R}$ replaces each q_i of a vector $\mathbf{q}$ by a corresponding probability distribution $\mathcal{R}q_i$.

We imagine the following operation of this scheme:

(1) The segmentation routines divide the original image into regions. Shape and texture descriptors are used to assign initial probabilities $p_i(\lambda)$ to appropriate labels λ for each region a_i.

(2) Information such as relative position and the nature of the boundary would be used to generate the compatibility coefficients r_{ij}.

(3) F would be iterated a few dozen times, say, to provide enhanced probabilities. If the result is unambiguous, interaction with other systems—perhaps using higher-level context more subtle than that expressible in the r_{ij}—could be involved in disambiguation, with the possibility of reinitiating F using a new set of probabilities.

We have already noted the similarity of the nonlinear probabilistic model to the S-RETIC, but with the emphasis on 'proper labelling' rather than on 'mode consensus'. Reviewing the Dev model of segmentation on prewired features in this light, we may note that it could be used for convergence to a sloping surface as well as for segmentation into regions of constant disparity. In fact, as Rosenfeld *et al.* [1975] note, their scheme can be used—as Dev has already done—to allow 'clusters' of low-level features having compatible labels to reinforce one another. For example, if the features are line orientations, we could use such a scheme to 'reinforce' those features which line up with their neighbors.

In conclusion, it seems that *cooperative computation*—a multi-level organization for problem-solving using many diverse, cooperating sources of knowledge, to use the title of the paper by Erman and Lesser [1975]—will provide the proper paradigm not only for the bottom-up and top-down approaches to brain theory (it seems to provide the right language for analyzing the neurological data of Luria [1973]), but also for artificial intelligence—a field which will provide many concepts for brain theory.

REFERENCES

1. R. P. Abelson, "The Structure of Belief Systems", Schank and Colby, 1973, 287–339.

2. M. A. Arbib, "Cognition—a Cybernetic Approach", Chapter 13 of *Cognition: a Multiple View* (Paul L. Garvin, ed.), Spartan Books, Washington, DC, 1970, 331–348.

3. ———, *The Metaphorical Brain*, Wiley-Interscience, New York, 1972.

4. ———, "Artificial Intelligence and Brain Theory: Unities and Diversities", *Annals of Biomedical Engineering*, 3 (1975), 238–274.

5. M. A. Arbib, C. C. Boylls, and P. Dev, "Neural Models of Spatial Perception and the Control of Movement", *Kybernetik und Bionik / Cybernetics and Bionics* (W. D. Keidel, W. Handler, M. Spreng, eds.), R. Oldenbourg, 1974, 216–231.

6. H. B. Barlow, C. Blakemore, and J. D. Pettigrew, "The Neural Mechanism of Binocular Depth Discrimination", *J. Physiol.*, 193 (1967), 327–42.

7. F. C. Bartlett, *Remembering*, Cambridge University Press, 1932.

8. V. Braitenberg and N. Onesto, "The Cerebellar Cortex as a Timing Organ", *Congress, Inst. Medicina Cibernetica*, Naples, 1960, 239–255.

9. B. Bruce and C. F. Schmidt, "Episode Understanding and Belief Guided Parsing", Association for Computational Linguistics Meeting, Amherst, MA, July 26–27, 1974.

10. P. Burt, "Computer Simulation of a Dynamic Visual Perception Model", *Int. J. Man-Machine Studies*, 7 (1975), 529–546. [See also P. Burt, *Stimulus Organizing*

154   *Michael A. Arbib*

Processes in Stereopsis and Motion Perception, COINS Technical Report 76-15, University of Massachusetts at Amherst, September 1976.]

11. P. Dev, "Segmentation Processes in Visual Perception: a Cooperative Neural Model", *Int. J. Man-Machine Studies*, **7** (1975), 511-528.

12. R. L. Didday, "The Simulation & Modelling of Distributed Information Processing in the Frog Visual System", Ph.D. Thesis, Stanford University, 1970.

13. R. L. Didday and M. A. Arbib, "Eye Movements and Visual Perception: a "Two Visual System" Model", *Int. J. Man-Machine Studies*, **7** (1975), 547-569.

14. L. Erman and V. Lesser, "A Multi-Level Organization for Problem-Solving Using Many, Diverse, Cooperating Sources of Knowledge", *Proc. 4th Intl. Joint Conf. Artificial Intelligence*, 1975.

15. H. G. Furth, *Piaget and Knowledge*, Prentice-Hall, Englewood Cliffs, NJ, 1969.

16. E. Goffman, *Frame Analysis: An Essay on the Organization of Experience*, Harper Colophon Books, N. Y, 1974.

17. A. R. Hanson, E. M. Riseman, and P. Nagin, "Region Growing in Textured Outdoor Scenes", *Proc. 3rd Milwaukee Symposium on Automatic Computation and Control*, 1975, 407-417.

18. B. Julesz, *Foundations of Cyclopean Perception*, University of Chicago Press, Chicago, 1971.

19. W. L. Kilmer, W. S. McCulloch, and J. Blum, "A Model of the Vertebrate Central Command System", *Int. J. Man-Machine Studies*, **1** (1969), 279-309.

20. J. Y. Lettvin *et al.*, "Two Remarks on the Visual System of the Frog", *Sensory Communication* (W. Rosenblith, ed.), M.I.T. Press, 1961.

21. A. R. Luria, *The Working Brain*, Penguin Books, 1973.

22. W. S. McCulloch, "A Heterarchy of Values Determined by the Topology of Nervous Nets", *Bull. Math. Biophys.*, **11** (1949), 89-93.

23. M. L. Minsky, "A Framework for Representing Knowledge", *The Psychology of Computer Vision* (P. H. Winston, ed.), McGraw-Hill, New York, 1975.

24. F. S. Montalvo, "Consensus vs. Competition in Neural Networks", *Int. J. Man-Machine Studies*, **7** (1975), 333-346.

25. J. D. Pettigrew, T. Nikara, and P. O. Bishop, "Binocular Interaction on Single Units in Cat Striate Cortex", *Exp. Brain Res.*, **6** (1968), 391-410.

26. W. H. Pitts and W. S. McCulloch, "How We Know Universals: The Perception of Auditory and Visual Forms", *Bull. Math. Biophys.*, **9** (1947), 127-147.

27. A. Rosenfeld, R. A. Hummel, and S. W. Zucker, "Scene Labelling by Relaxation Operations", *IEEE Trans. Systems, Man, and Cybernetics*, **SMC-6** (1976), 420-433.

28. R. C. Schank, "Identification of Conceptualizations Underlying Natural Language", Schank & Colby, 1973, 187-247.

29. R. C. Schank and K. M. Colby (eds.), *Computer Models of Thought and Language*, W. H. Freeman, San Francisco, 1973.

30. J. R. Searle, "What is a Speech Act?", *Philosophy in America* (Max Black, ed.), Allen & Unwin, London, 1965, 221–239.

31. O. L. Selfridge, "Pandemonium: a Paradigm for Learning", Mechanization of Thought Processes, H.M.S.O., London, 1959, 513–526.

32. H. R. Wilson and J. D. Cowan, "A Mathematical Theory of the functional dynamics of cortical and thalamic nervous tissue", *Kybernetik*, **13** (1973), 55–80.

33. T. Winograd, *Understanding Natural Language*, Academic Press, New York, 1972.

34. ———, "A Procedural Model of Language Understanding", Schank & Colby, 1973, 152–186.

MATHEMATICAL MODELS FOR
CELLULAR BEHAVIOR*†

Lee A. Segel

1. INTRODUCTION

Triumphant in their search for basic molecular mechanisms, many biologists are turning to the examination of how such mechanisms guide cellular behavior. Under intensive investigation is the rich variety of behavior exhibited by single cells. In addition, much research is concerned with collective behavior, particularly in the contexts of neurology and developmental biology.

The excitement of discovering the basic ingredients of cellular conduct must be left to biologists, but mathematics is indispensable in fully analyzing the consequences of this conduct. Thus, the present paper deals with the construction and analysis of

*The National Science Foundation supported the portion of the work that was done at Rensselaer Polytechnic Institute under Grant MCS 76-07492

†This paper is dedicated to the memory of my friend Julius L. Jackson. His untimely death occurred just before we were to renew our collaboration on problems of the type discussed here. Science moves on almost heedless of individual fates, but one never fully recovers from the loss of a warm and stimulating colleague.

mathematical models for certain types of cellular behavior that have captured my interest and that of several other applied mathematicians.

It might legitimately be asked whether it is not better for biologists themselves to construct the models and for mathematicians to enter only at the next stage. This in fact frequently happens, but there are many cases where mathematicians have been deeply involved in the creation of appropriate models. A model is a representation of reality so that there is much freedom of choice in its construction. Selection of a representation that can without undue labor yield significant new understanding requires not only a thorough understanding of the facts but also a knowledge of available theoretical concepts and analytic tools. "Which is a more appropriate representation of cellular motion—a biased random walk or a continuous diffusion process supplemented by an additional gradient-dependent flux?" The present article is partially concerned with this question, which is a natural one for an applied mathematician. But few biologists have been able to spare time from their more central undertakings to become thoroughly familiar with the concepts that would lead one to such a query. This illustrates why mathematicians have a role to play in biological modelling.

As will be seen, the mathematics involved in the analysis of new biological models is often very simple. Indeed, at first the major contribution is to select a model that will yield striking results in a simple manner—for often only such results have a strong impact on the biological community. Later in the development of the theory, however, more dignified mathematics may be involved. This too will be briefly illustrated.

We shall begin our discussion with a problem involving slime mold amoebae, organisms which secrete a chemical attractant that can have strong effects on their collective motion. In this case we shall give a considerable amount of biological background in order to provide a flavor of what is required in constructing an appropriate model. The organisms will be regarded as continuously distributed so that the model will take the form of a set of partial differential equations. We shall explicitly present the very simple stability analysis that already yields major qualitative features of the cells' characteristic aggregating movement. We shall then briefly mention some later work that looks more deeply into the subtleties of the matter.

We next turn to a problem concerning swimming bacteria that are attracted to certain chemicals. The model is of a random walk type. Appropriate limits will be shown to yield useful differential equations. (The material here is new.)

We shall briefly mention recent work that goes beyond a descriptive approach to the bacterial motion and takes into account the chemical interaction between molecules of attracting chemical and the large molecules that constitute the bacteria's receptor system. Here too the model takes the form of a set of partial differential equations, for which the earlier random walk approach provides a basis.

The paper concludes with an overview of the various types of models that have been used to study cellular behavior.

2. AGGREGATION OF SLIME MOLD AMOEBAE

Some biological facts. We present here some biological facts that provide necessary background to the construction of our mathematical model. We shall restrict our remarks to the most studied species of cellular slime molds, *Dictyostelium discoideum*. There are several other species, with corresponding differences of behavior (Bonner 1967)*.

One can begin a description of the life cycle of the slime mold amoebae at the spore stage, where each amoeba is dormant with a protective covering. When conditions are favorable, an amoeba emerges from its spore. Of the order of 10 micrometers (10^{-3} cm) across, the amoebae are rather shapeless one-celled organisms that move by extending contractile portions of themselves (pseudopods).

The natural habitat of the amoebae is soil or dung. An im-

---

*From this point until the three paragraphs headed "Interpretation of the analysis", the material is taken with only minor changes from Lin and Segel (1974). (Permission of the publishers has been secured.) The basic source of the theoretical content of this material is an article by Keller and Segel (1970).

portant element of the food chain on earth, they feed on bacteria by engulfing them. If food is plentiful, the amoebae continually feed and multiply by mitosis (dividing in two). If the food supply becomes exhausted there is an *interphase period* of random and somewhat feeble movement, where the amoebae are more or less evenly distributed over the area available to them. During this period, the disappearance of the food supply triggers certain changes in the amoebae. The details of these internal changes are not known, but there is no difficulty in observing the striking external phenomenon that results. After a few hours, the amoebae begin to aggregate into a number of collection points. Typically, these are more or less regularly distributed, with a spacing of a few hundred micrometers.

After aggregation has been completed, the amoebae that have collected at a given point (ranging in numbers from a few in laboratory experiments, up to 200,000) form a multicelled slug. This moves as a unit, although the formerly free-living amoebae retain their cell walls within the slug. Then the slug stops and erects a stalk, on top of which is a roundish container of spores. The cycle is thus completed.

What is responsible for the organized aggregation of the amoebae? This is an important question, for "purposeful" movements occur frequently in developmental processes. Usually, these movements take place rather inaccessibly within a developing organism, but the cells of the slime mold amoebae will obligingly perform in a laboratory dish so that it is easy to examine them and to experiment with them.

It has been discovered that the amoebae move preferentially toward relatively high concentrations of a chemical which they themselves secrete. In some species the attractant has recently been identified as cyclic 3', 5'-adenosine monophosphate (AMP), a chemical with many important biological functions. It is also known that a given quantity of attractant loses its potency in a matter of minutes. This has been traced to the activity of an enzyme that alters the nature of the AMP.

Presumably, aggregation is caused by the fact that the amoebae move up a gradient of attractant, but what determines the time of onset? What determines the spacing of aggregation centers? Can one quantify the process? A mathematical model is needed to answer such questions. We shall proceed to devise the simplest model that could reasonably be supposed to bear on the circumstances. If analysis of this model is encouraging, one can add detail later.

Formulation of a mathematical model. Since the distance between amoebae is small compared to the typical distance between aggregation centers, we shall employ a continuum model. Suppose that the aggregation takes place in the (x, y) plane. For simplicity we shall assume that no quantities change with y, so that only x variation need be considered. This assumption of unidimensionality is not at all essential, but it makes exposition easier.

Let $a(x, t)$ denote the number of amoebae per unit area at position x and time t. Consider the amoebae located in the region $x_0 \leqslant x \leqslant x_0 + \Delta x$, where Δx is an arbitrary number (not necessarily small). We shall now write a *general balance law* for this region; this states that the rate of change of the net amount of a in the region equals the rate at which a flows across the boundary, plus the net rate of creation of a within the region. In the present case, a stands for amoebae, and the net creation of amoebae is equal to the excess of births over deaths. But the balance law is "general" because it applies to any substance whatever. For how else can a substance appear in the region except by creation or by flow across its boundaries?

To proceed, we must define the *flux density* $J(x_1, t)$. This gives the net rate per unit length at which a crosses the line $x = x_1$. Also, J is defined to be positive if there are more amoebae crossing in the direction of x increasing than in the opposite direction. The term $Q(x, t)$ is the net rate per unit area at which a is being created. The desired balance law can now be written. Considering the rectangle of Figure 1 we obtain

$$\frac{\partial}{\partial t} \int_{x_0}^{x_0 + \Delta x} a(x, t)dx = J(x_0, t) - J(x_0 + \Delta x, t)$$

$$+ \int_{x_0}^{x_0 + \Delta x} Q(x, t)dx. \qquad (1)$$

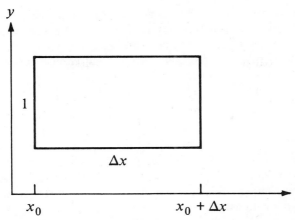

FIG. 1. The balance law (3) considers the rate of change of mass in a rectangle of length Δx and unit width, in the case where y variation is absent.

It is convenient to use the integral mean value theorem to write

$$\frac{\partial}{\partial t} \left[a(x_1, t)\Delta x \right] = J(x_0, t) - J(x_0 + \Delta x, t) + Q(x_2, t)\Delta x;$$

$$x_0 \leqslant x_1 \leqslant x_0 + \Delta x, \quad x_0 \leqslant x_2 \leqslant x_0 + \Delta x. \quad (2)$$

We divide by Δx and then take the limit as $\Delta x \to 0$, to obtain the *general balance law in differential equation form*,

$$\frac{\partial a}{\partial t}(x_0, t) = -\frac{\partial J}{\partial x}(x_0, t) + Q(x_0, t), \quad x_0 \text{ arbitrary.} \quad (3)$$

In the case of the amoebae, reproduction can be ignored because there is little or none of it in the absence of food. Deaths can also be ignored, since there are few during the time interval of interest. Thus we take

$$Q \equiv 0. \quad (4)$$

To obtain an expression for the flux J, let us first consider a situation when attractant is absent. Then the amoebae appear to move about randomly. Owing to such "diffusionlike" movement, a concentration of amoebae tends to disperse. Thus there is a random flux J, from regions of high amoeba concentration to regions of low concentration. The magnitude of the flux at x seems to depend on the concentration difference between x and nearby points. We characterize this difference by $\partial a / \partial x$ (the simplest choice) and make the hypothesis that

$$J_r(x, t) = F\left[\frac{\partial a(x, t)}{\partial x}\right] \quad (5)$$

for some function F. Now when $a \equiv$ constant, $J = 0$—for in random motion there will be as many amoebae moving to the left as to the right. In this case $\partial a / \partial x \equiv 0$. Therefore, given (5), it is only sensible to assume that

$$F(0) = 0. \quad (6)$$

Thus the graph of F must have an appearance such as that depicted in Figure 2. For sufficiently small values of s, we can approximate the graph by a straight line. Calling the slope of the line $-\mu$, we obtain $F(s) = -\mu s$ as the simplest reasonable assumption concerning F; i.e.,

$$J_r(x, t) = -\mu \frac{\partial a(x, t)}{\partial x}. \quad (7)$$

Combining (3), (4), and (7), we obtain

$$\frac{\partial a}{\partial t} = \frac{\partial}{\partial x}\left(\mu \frac{\partial a}{\partial x}\right). \quad (8)$$

Equation (8) is the *diffusion equation*, which is used to model the concentration variation of any kind of randomly moving set of particles, for example, smoke particles in air. The constant μ, which governs the vigor of the random movement, is generally called the *diffusivity*. Here we call μ the *motility*, giving precise meaning to a biological term that is often used only in a qualitative manner. Application of a simple diffusion theory to provide a quantitative assay for motility is discussed by Segel, Chet, and Hennis (1977).

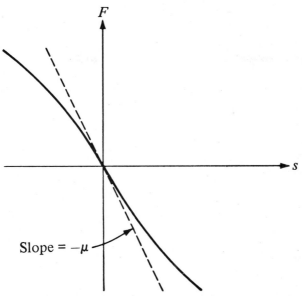

Fig. 2. Possible plot of the function F defined in (5). The graph provides negative values for $s > 0$ because there should be a leftward (negative) flux if amoeba density is higher at larger values of x. Similar reasoning explains the positivity for $s < 0$.

Equation (8) was obtained under the assumption that no attractant was present. To account for *chemotaxis*, which is directional motion induced by variations in chemical concentration, we add to J_r of (7) an additional contribution J_c. Let $p(x, t)$ be the density of the attractant. Arguing as before, we pass from an initial assumption that J_c is a function of the attractant gradient $\partial p / \partial x$ to the assumption that J_c is proportional to the attractant gradient, at least for small values of the gradient. For a given gradient, if the amoebae density is twice as great, the net flux should be twice as great. The proportionality factor should thus be a multiple of a. We are led to the hypothesis that

$$J_c = \chi a \frac{\partial p}{\partial x} . \qquad (9)$$

The factor χ measures the strength of chemotaxis. Note that in contrast to (7), there is no negative sign in (9). This is because amoebae tend to move *toward* attractant concentrations (and away from amoeba concentrations).

Assuming that the total flux J in (3) is the sum of the random contribution J_r and the chemotactic contribution J_c, we are led to our final equation for the change in amoeba density:

$$\frac{\partial a}{\partial t} = \frac{\partial}{\partial x}\left(\mu\,\frac{\partial a}{\partial x} - \chi a\,\frac{\partial p}{\partial x}\right). \tag{10}$$

We shall take μ and χ to be positive constants. It is not difficult to take into account a variation with p which is probably present, but this would only change some details of the analysis to come. Note that even with constant μ and χ, (10) contains a (quadratic) nonlinear term $\chi a(\partial p/\partial x)$, because this term is proportional to the product of two unknown functions, a and $\partial p/\partial x$.

We also need an equation for the attractant density p. This will be of the general form (3):

$$\frac{\partial p}{\partial t} = -\frac{\partial J_a}{\partial x} + Q_a.$$

(The subscript "a" refers to attractant.) The random motion of the attractant molecules will be modeled by a proportionality of flux to gradient, as in (7):

$$J_a = -D\,\frac{\partial p}{\partial x}.$$

The net creation term Q_a has a positive contribution fa as a result of the secretion of attractant by the amoebae. Here f is the rate of secretion per unit amoebae density. What of the decay in attractant activity? We take the rate of decay (as in radioactivity or some other spontaneous process) to be proportional to the amount of attractant present,* via the constant k. Thus $Q_a = fa - kp$,

---

*As we stated above, decay is actually due to the action of an enzyme. This can be modelled to some extent by making k a certain function of p, but the essentials of the analysis are thereby unchanged. A more comprehensive approach to the enzyme action will be mentioned below.

and the desired equation for $\partial p / \partial t$ is

$$\frac{\partial p}{\partial t} = fa - kp + D \frac{\partial^2 p}{\partial x^2}. \tag{11}$$

We shall take f, k, and D to be positive constants.

Partial differential equations (10) and (11), for the two unknown functions $a(x, t)$ and $p(x, t)$, provide the basic formulation of our problem.

An exact solution: the uniform state. It is very easy to find an exact solution of (10) and (11). This is the *uniform solution*

$$a = a_0, \quad p = p_0, \tag{12}$$

where a_0 and p_0 are constants. [When (12) holds, the system is said to be in the *uniform state*.] Upon substitution into (10) and (11), we find that (12) will indeed provide a solution if

$$fa_0 = kp_0. \tag{13}$$

Equation (13) is physically reasonable. It says that in the uniform state the secretion rate of attractant must be exactly balanced by the decay rate.

Analysis of aggregation onset as an instability. We identify the uniform state with the interphase period prior to aggregation. We also model the onset of aggregation as the *breakdown of the uniform state due to the growth of inevitable small disturbances* to amoeba and attractant density. That is, we identify the onset of aggregation with the sort of *instability of the uniform state* whose investigation forms a classical part of applied mathematics. The instability is presumed to occur because of changes, during interphase, of the parameters μ, χ, and f that characterize amoeba behavior.

The idea behind instability theory is this. Suppose that at some initial time, the state of the system is slightly disturbed from the uniform state. (Suppose that there is a local clumping of amoebae, for example, and an accompanying local concentration of attractant.) Will the small disturbance tend to disappear with the passage of time or will it become more intense? In the former case we say that the uniform state is *stable to small disturbances*, in the latter *unstable*. Unstable states will not be observed, for disturbances are inevitable. In the case of instability, they will grow, so the uniform state will be replaced by some other state.

To perform a stability analysis, we introduce the variables a' and p' by the definitions

$$a(x, t) = a_0 + a'(x, t), \quad p(x, t) = p_0 + p'(x, t). \quad (14)$$

Here $a' = a - a_0$, for example, measures departure from uniformity; therefore, it can be identified with the disturbance in amoeba density.

To obtain equations for a' and p', we substitute (14) into (10) and (11). From the former we obtain

$$\frac{\partial a'}{\partial t} = \mu \frac{\partial^2 a'}{\partial x^2} - \chi \left[(a_0 + a') \frac{\partial^2 p'}{\partial x^2} + \frac{\partial a'}{\partial x'} \frac{\partial p'}{\partial x'} \right]. \quad (15)$$

This equation is nonlinear, owing to the presence of the quadratic terms $a'(\partial^2 p'/\partial x^2)$ and $(\partial a'/\partial x')(\partial p'/\partial x')$. We shall assume that the disturbances (and their derivatives) are small, in which case we call a' and p' *perturbations*. The perturbations are involved quadratically in some terms and linearly in others. Products of two small terms should be negligible in comparison with the other terms of (15), which contain but a single perturbation function. We thus *linearize* the equation by *deleting all nonlinear terms*. We obtain

$$\frac{\partial a'}{\partial t} = \mu \frac{\partial^2 a'}{\partial x^2} - \chi a_0 \frac{\partial^2 p'}{\partial x^2} \quad (16)$$

as the linearized version of (10). As for (11), upon substituting (14) and employing (13), we obtain

$$\frac{\partial p'}{\partial t} = fa' - kp' + D \frac{\partial^2 p'}{\partial x^2}, \quad (17)$$

which is already linear.

We are faced with a pair of linear partial differential equations with constant coefficients. We guess that there are solutions of the form

$$a' = C_1 \sin qx \, e^{\sigma t}, \quad p' = C_2 \sin qx \, e^{\sigma t}, \qquad (18)$$

where C_1 and C_2 are constants.* It is easily seen that there are indeed solutions of the form (18), provided that

$$(\sigma + \mu q^2)C_1 - \chi a_0 q^2 C_2 = 0, \qquad (19a)$$

$$- f C_1 + (\sigma + k + D q^2)C_2 = 0. \qquad (19b)$$

This system of algebraic equations has the trivial solution $C_1 = C_2 = 0$. [From (18) we see that we have merely verified that it is possible to have an identically zero perturbation of an exact solution.] For a nontrivial solution, the determinant of the coefficients must vanish. We thus obtain the following quadratic equation for σ:

$$\sigma^2 + b\sigma + c = 0;$$

$$b \equiv k + (\mu + D)q^2, \quad c \equiv \mu q^2(k + Dq^2) - \chi a_0 f q^2. \quad (20)$$

The quadratic equation can be shown to have real roots. To ensure stability, both roots must be negative, so that the exponential factor $\exp(\sigma t)$ brings about decay of the perturbations. It is not difficult to show that a necessary and sufficient condition for stability is $c > 0$, which requires that

$$\chi a_0 f < \mu(k + Dq^2), \quad q \neq 0. \qquad (21)$$

From (18), $2\pi/q$ is the wavelength of the perturbation under investigation. Since μ and k are positive, the right side of (21) decreases monotonically as q decreases, toward a greatest lower bound of μk. Therefore, the longer the wavelength $2\pi/q$, the more

---

*A cosine dependence in (18) will yield exactly the same results. Also, more general disturbances can be obtained by the superposition of sinusoidal solutions, using Fourier analysis.

"dangerous" the perturbation [for (21) is more easily violated]. We conclude that *instability* is possible whenever

$$\frac{\chi a_0 f}{\mu k} > 1, \tag{22}$$

for then (21) is violated for a range of sufficiently small values of q^2.

Interpretation of the analysis. The view of aggregation that emerges is this. During the beginning of the interphase period, inequality (22) does not hold and the uniform state is stable. Triggered by the stimulus of starvation, the various parameters gradually change. Eventually, (22) is satisfied, and aggregation commences.

The instability criterion (22) has the following interpretation. Suppose that there is a concentration of amoebae and attractant at some point. Random "diffusion" of amoebae with motility μ tends to disperse this concentration, as does the attractant decay, whose strength is measured by k. Larger μ and k means larger stabilizing effects. It is therefore appropriate that increases in μ and k mean that the instability criterion (22) is more difficult to satisfy.

By contrast, a local concentration of attractant tends to draw amoebae toward it by chemotaxis (strength χ). Also, a concentration of amoebae will provide a corresponding increase of attractant concentration because of the higher local concentration of secretion sources. The strength of this effect is measured by $a_0 f$. This explains the appearance of the destabilizing factors $a_0 f$ and χ in the numerator of (22). Instability ensues if destabilizing effects outweigh stabilizing.

In terms of the present model, the size of the "aggregation territory" must be identified with the wavelength λ_{max} that makes the growth rate σ a maximum. We have seen that just at the onset of instability $\lambda_{max} = \infty$. In general one finds (see Segel 1971) by a regular perturbation expansion for the largest root of (20) that

$$\lambda_{max} = 2\pi \left(\frac{2D}{k\Delta} \right) [1 + O(\Delta)]. \tag{23}$$

Here

$$\Delta \equiv \frac{\chi a_0 f}{\mu k} - 1$$

is a measure of how far the critical condition for the onset of instability has been exceeded.

The model considered here has been extended by taking into account the fact that attractant decay is due to the action of an enzyme. Two new equations supplement (10) and (11). The quadratic (20) becomes a quartic of the form

$$\epsilon\beta\sigma^4 + (c_1 q^2 + c_2)\sigma^3 + (c_3 q^4 + c_4 q^2 + c_5)\sigma^2$$

$$+ (c_6 q^6 + c_7 q^4 + c_8 q^2 + c_9)\sigma$$

$$+ (c_{10} q^8 + c_{11} q^6 + c_{12} q^4 + c_{13} q^2) = 0. \qquad (24)$$

The coefficients c_i are functions of various parameters. In particular they are polynomials of degree at most two in ϵ and β. Little was known of the magnitudes of the various parameters at the time the extended model was examined, so analysis of the algebraic equation (24) (using ideas of singular perturbation theory) was restricted to a search for new types of qualitative behaviour. For certain parameter ranges a finite "most dangerous" wavelength was predicted to occur at the onset of instability. In other cases, it was shown that aggregation can commence in a pulsatile manner. Details of the analysis and comparison with experiment can be found in Segel and Stoeckly (1972).

Nanjundiah (1973) considered various extensions of the work reported so far. Azimuthal instabilities (as an explanation for a "streaming" pattern sometimes found) and nonlinear effects were among the topics treated. In addition, references can be found in Nanjundiah's paper to several other noteworthy theoretical papers on aggregation by members of the theoretical biology group at the University of Chicago.

3. MODELLING BACTERIAL CHEMOTAXIS

Observations concerning bacterial chemotaxis. Many bacteria are motile (capable of movement) by virtue of whiplike flagellae whose motion propels them through their fluid environment. Such bacteria seem invariably repelled or attracted by one or more chemicals. This chemotaxis has recently been an object of intensive study, since it provides a simple model for the reception of an environmental signal and its transduction into appropriate action. It is hoped that understanding of rudimentary bacterial "nervous systems" will increase understanding of true nervous systems in higher organisms.

We now list a few of the salient facts about bacterial chemotaxis in the most studied species *Escherichia coli* and *Salmonella*. Further information can be obtained from the review by Koshland (1974) and the references cited therein. (a) In the absence of chemical gradients, bacterial motion consists of more or less straight paths interspersed with sudden tumbles. (b) The probability per unit time of a tumble is roughly constant. After tumbling a bacterium begins to move in a new direction that appears randomly selected, but there is some bias toward forward motion. (c) Tumbling decreases noticeably when a bacterium moves toward an attractant, but seems only slightly affected when the bacterium moves away from an attractant. (d) A bacterium does not sense the presence of a gradient by instantaneously comparing concentrations at different points on its outer membrane. Rather there is a crude "memory capacity" that allows a bacterium to compare present concentration with that in the past, and to persist or to tumble according to the results of this comparison.

We shall shortly present a model for bacterial chemotaxis based on the notion of biased random walk. Keller and Segel (1971) made an early attempt to formulate a random walk model for chemotaxis in amoebae and bacteria. Since then, the phenomena have become more widely known, more facts have been ascertained, and the number of papers has grown. Recent examples are articles by Stroock (1974) and Nossal and Weiss (1974). Here we pay particular attention to issues involved in passing from the difference equations of the random walk model to appropriate limiting differential equations.* To provide necessary contrast and perspective, we shall preface our main discussion by a review of the classical random walk model for Brownian motion and its diffusion limit.

Brownian motion.   In Brownian motion, dust particles (for example) move about under the influence of innumerable jostles by tiny air molecules. A particle's center of mass traverses a contorted path, as indicated schematically in Figure 3. We wish to obtain a more manageable description of the phenomenon. To this end we select a time interval τ_1, where τ_1 is long compared to the time interval between molecule-particle collisions, and we define a step initiated at time t as the displacement of the particle in the time interval $(t, t + \tau_1)$. That is if $\mathbf{x}(t)$ is the position of the particle's center of mass at time t, the length $\delta(t, \tau_1)$ of this step will be given by

$$\delta(t, \tau_1) = |\mathbf{x}(t + \tau_1) - \mathbf{x}(\tau_1)|.$$

The motion of the particle can now be idealized as a succession of steps, each taking a time interval τ_1.

Successive molecular collisions with the dust particle will be correlated, for nearby molecules influence each other's motion. But since a very large number of collisions occur during the time interval τ_1, we can safely make the important simplifying assumption that successive *steps* are *uncorrelated*.

Suppose we select a shorter step duration time τ_2 that still is large compared to the time between collisions. We note for later reference that the motion of the particle will be better resolved so that a higher "apparent speed" will be observed. To see this, suppose for example that $\tau_2 = \frac{1}{2}\tau_1$. If we estimate the speed by observing the distance "stepped" during the time interval (t, τ_1), then we have from Figure 3

$$\frac{\delta(t, \tau_1)}{\tau_1} < \frac{\delta(t, \tau_2) + \delta(t + \tau_2, \tau_2)}{2\tau_2} . \tag{25}$$

---

*The author is grateful to M. D. Kruskal for helping him clarify these issues.

FIG. 3. The continuous line represents the path of a Brownian particle. Points A, B, and C denote the particle position at times t, $t + \tau_2$ and $t + \tau_1$, $\tau_1 \equiv 2\tau_2$. Choosing the smaller value τ_2 for a step duration gives rise to a higher apparent speed. A step duration $\tau_3 < \tau_2$ would yield a still higher apparent speed, etc.

Viewing our model in perspective, we see there are two broad restrictions on the time intervals τ_i that we might select. First, our time interval cannot be too long, otherwise our view of the Brownian motion is so crude that no interesting details can be discerned. Thus the τ_i must be small compared to the time scale of "interesting" phenomena, for example, the time it takes for smoke particles to diffuse a few meters from a smoke stack. Second, τ_i must be large compared to the time between molecule-particle collisions (typically 10^{-10} secs); otherwise successive steps would be correlated and the whole basis for the argument to come would be destroyed*.

To avoid some technical complications, we shall simplify our model still further. Let us project the motion on a line and assume that the particle moves a fixed distance Δ_i during each time interval τ_i. We select Δ_i to be the average projected distance moved in the time interval τ_i. We choose the x-axis along the given

---

*It is tempting to model the effect of turbulent fluid flow by a Brownian-like random fluid motion, but grave difficulties generally ensue because local turbulent effects are correlated for times comparable to the time scale of large-scale changes in the flow. See Corrsin (1974) for a recent discussion of this matter.

line and we denote by **u** the unit vector along this axis, so that

$$\Delta_i = \lim_{T \to \infty} \frac{1}{2T} \int_{-T}^{T} |[\mathbf{x}(\tau_i + t) - \mathbf{x}(\tau_i)] \cdot \mathbf{u}| dt. \qquad (26)$$

The non-correlation between successive steps is reflected in this simplest of models by the hypothesis that each step of the particle along the x-axis is taken with equal probability to the left or to the right.

We are now ready for the standard presentation of one dimensional random walk and its diffusion limit. This topic is not central here and is covered in many texts (for example, in Lin and Segel 1974, Chapter 3), so that we only outline the arguments. A point or two of special interest will be given a little more fully however.

Select a time interval τ_i. Let x denote position along the line, and let $n(x, t)$ denote the expected number of particles at point x, time t. Particles at x at time $t + \tau_i$ must, at time t, have been located either at $x + \Delta_i$ or $x - \Delta_i$ and then have moved appropriately left or right with probability $1/2$. (Here is where the non-correlation between steps is used.) Consequently

$$n(x, t + \tau_i) = \frac{1}{2} n(x - \Delta_i, t) + \frac{1}{2} n(x + \Delta_i, t). \qquad (27)$$

Consider a sequence of decreasing time intervals τ_i and corresponding (decreasing) step lengths Δ_i. Making the dependence on τ_i and Δ_i explicit in our notation, we see that (under weak assumptions)

$$n(x, t, \tau_i, \Delta_i) = n(x, t, 0, 0) + O(\tau_i) + O(\Delta_i). \qquad (28)$$

The time between collisions is so short compared to the time scale of macroscopic phenomena that τ_i can be taken to be very small. It is therefore reasonable to presume that the $O(\tau_i)$ and $O(\Delta_i)$ terms are negligible in (28), so that $n(x, t, \tau_i, \Delta_i)$ can be approximated by $n(x, t, 0, 0)$. This approximation may be written

$$n(x, t, \tau_i, \Delta_i) \approx \lim_{\tau_i, \Delta_i \to 0} n(x, t, \tau_i, \Delta_i). \qquad (29)$$

In spite of the fact that a limit as $\tau_i \to 0$ appears in (29) it should nonetheless be clear that this approximation does not require us to abandon the requirement that τ_i remain large compared to the average time between particle-molecule collisions.

For any fixed initial number of particles, the number of particles expected on a given location approaches zero as the number of locations increases. A quantity that remains finite as $\Delta_i \to 0$ is obtained by introducing the number of particles per unit length $b(x, t)$, where

$$b(x, t) = n(x, t)/\Delta_i.$$

It can be shown that the root mean squared displacement in time t, for the random walk model we have introduced, is approximately $t\Delta_i^2/\tau_i$. To keep the bulk of particles under scrutiny we must keep this fixed as Δ_i and $\tau_i \to 0$. Thus a more accurate description of the limit in (29) can be written as follows:

$$\lim(\tau_i \to 0, \Delta_i \to 0, D \text{ fixed}), \quad \text{where } D = \frac{\Delta_i^2}{2\tau_i}. \tag{30}$$

When applied to the governing difference equation (27), a Taylor expansion shows that this limit gives the diffusion equation

$$\frac{\partial b}{\partial t} = D \frac{\partial^2 b}{\partial x^2}.$$

A final observation must be made for later reference. The limit (30) requires

$$\frac{\Delta_i}{\tau_i} = \frac{2D}{\Delta_i} \to \infty.$$

Our earlier remarks lead us to expect this continual increase in average speed with decreased τ_i (and therefore with increased resolution of the Brownian path).

Comparison of chemotactic and Brownian motion. Turning now to bacterial chemotaxis, we first observe a strong similarity between the motion of an individual bacterium and that of an individual dust particle in Brownian motion. Both particles move in a more or less straight line for a period and then make a relatively radical change of direction. In both cases this change of direction is due to molecular activity that "in principle" is computable—exterior molecular activity for the dust specks, interior for the bacteria. But in both cases the computation presents insuperable practical difficulties and a hypothesis of randomness is in order. We shall employ a one-dimensional representation of bacterial motion, as we did in the case of Brownian movement.

A major contrast between Brownian motion and bacterial chemotaxis is connected with the fact that the trajectory of a particle between radical direction changes is impossible to observe in the former case and is readily observable in the latter. In fact the linear speed v of a bacterium can be obtained by microscopic observation of its motion between turns. We shall assume for simplicity (as is reasonable in many situations) that v is a constant. We expect this speed to be a parameter in models for bacterial motion; by contrast, we have seen that no finite particle speed appears in the simplest Brownian motion model and we have argued that this is reasonable.

Since we wish the speed to appear in our model, a "step" in bacterial motion should be taken as a fraction of the distance traversed between turns. Presumably it is some sort of transthreshold fluctuation in a local concentration of control molecules that brings about a sudden change in bacterial direction. A lower limit on step duration results from the necessity to insure that concentrations of such molecules are not correlated from one step to the next. Subject to this limitation, we can consider smaller and smaller step durations τ. Unlike the situation in models for Brownian motion, however, here the speed is independent of the choice of τ (as long as τ is a fraction of the average interval between turns). (See Figure 4.) Because a "step" is a small fraction of a straight segment in the particle's path, it is now appropriate to assign probabilities to a particle's tendency to persist or, alternatively, to choose a new direction.

Keeping the above remarks in mind, let us now formulate a one-dimensional probabilistic model for bacterial motion.

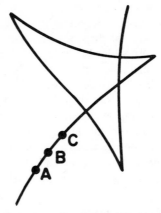

FIG. 4. The line represents the path of a chemotactic bacterium. Points A, B, and C denote the bacteria position at times t, $t + \tau$, $t + 2\tau$.

Bacterial chemotaxis as a random walk. We consider a one-dimensional situation in which particles (i.e., bacteria) move to the left or right along a line. Let x denote position along this line and let the superscripts "+" and "−" distinguish right- and left-moving particles respectively. At time t, a certain number $n^+(x, t)$ of particles will arrive at point x from the left. The number of particles that arrive at point x from the right at time t will be denoted by $n^-(x, t)$.*

Let $p^+(x, t)$ be the persistence probability for a particle arriving at x from the left at time t. That is, let p^+ be the probability that the particle continues from x in the positive direction. In addition, let $q^+(x, t)$ be the corresponding reversal probability. Since the particle must either persist or reverse we have

$$p^+(x, t) + q^+(x, t) = 1. \tag{31}$$

Similarly, we define persistence and reversal probabilities, p^- and q^-, for left-moving particles, where

$$p^-(x, t) + q^-(x, t) = 1. \tag{32}$$

---

*Sometimes we write $n^\pm(x, t; \Delta)$ to make explicit the dependence on Δ.

Suppose that the particle executes a random walk, subject to the above probabilities, with a fixed step length Δ. Let the duration of a step be a fixed time τ, so that the particle moves with velocity v given by

$$v = \Delta/\tau. \tag{33}$$

The motion of the particles is governed by the following equations:

$$n^+(x + \Delta, t + \tau) = n^+(x, t)p^+(x, t) + n^-(x, t)q^-(x, t),$$
$$\tag{34a}$$

$$n^-(x - \Delta, t + \tau) = n^+(x, t)q^+(x, t) + n^-(x, t)p^-(x, t).$$
$$\tag{34b}$$

Equation (34a) is based on the fact that the particles which arrive at $x + \Delta$ from the left, at time $t + \tau$, must have been located at point x at time t. The first term on the right side of (34a) gives the number of particles that reached $x + \Delta$ by arriving at x from the left and persisting. The second term gives the number that reached $x + \Delta$ by arriving at x from the right and reversing. Equation (34b) can be obtained by entirely similar reasoning.

For the moment, let us regard the probabilities $p^\pm$ and $q^\pm$ as given. Then equations (34a) and (34b), perhaps together with some boundary conditions, completely describe the evolution of an initial particle distribution providing that some initial condition is given. It might be postulated for example that the first step is either to the left or to the right with equal probability.

Limiting differential equations. Rather than attempt to contend with the initial value problem for (34) as it stands, we turn to the task of approximating (34) in the case of small Δ and τ. This will lead us to a set of partial differential equations from which general populational trends can be deduced.

We have pointed out that decreasing the step size Δ will leave the speed unaltered. But as Δ is decreased, there is a proportionately increasing number of steps between tumbles. The probability of tumbling in a single step should therefore approach zero like Δ as the step size decreases toward zero. In our one-dimensional model, a tumble corresponds to a reversal of direction. Thus, the appropriate limit in the present case is

$$\Delta \to 0, \quad \tau \to 0, \quad \Delta/\tau \equiv v \text{ fixed}, \quad q^\pm = O(\Delta). \quad \text{(35a, b, c, d)}$$

This limit should be compared to (30).

It is convenient to regard (34a) and (34b) as difference equations for $n^\pm(x, t; \Delta)$ in which there appears a small parameter Δ. [By (35c), the parameter τ can be regarded as an abbreviation for Δ/v.] We shall seek a solution in the form of an expansion in powers of Δ.

We thus write, for example,

$$n^\pm(x, t; \Delta) = n_0^\pm(x, t) + \Delta n_1^\pm(x, t) + \cdots. \quad (36)$$

Also, in keeping with (35d) we assume

$$q^\pm(x, t; \Delta) = \Delta q_1^\pm(x, t) + O(\Delta^2), \quad (37a)$$

$$p^\pm(x, t; \Delta) = 1 - \Delta q_1^\pm(x, t) + O(\Delta^2). \quad (37b)$$

Expansions of the following character must be made:

$$n^+(x + \Delta, t + \tau)$$

$$= n_0^+(x, t) + \Delta\left[n_1^+(x, t) + \frac{\partial n_0^+}{\partial x}(x, t) + \frac{1}{v}\frac{\partial n_0^+}{\partial t}(x, t) \right] + \cdots.$$

With these we find that (34a) and (34b) are identically satisfied at $O(1)$. Equating the $O(\Delta)$ terms to zero we obtain

$$\frac{\partial n_0^+}{\partial x} + \frac{1}{v}\frac{\partial n_0^+}{\partial t} = -n_0^+ q_1^+ + n_0^- q_1^-, \quad (38a)$$

$$-\frac{\partial n_0^-}{\partial x} + \frac{1}{v}\frac{\partial n_0^-}{\partial t} = n_0^+ q_1^+ - n_0^- q_1^-. \quad (38b)$$

To interpret these equations, and for later use, we introduce the notation

$$\sigma^{\pm} \equiv \frac{1}{\tau} q^{\pm} \approx v q_1^{\pm}; \quad \frac{n_0^{\pm}}{\Delta} \equiv b^{\pm}. \tag{39}$$

The quantities σ^+ and σ^- give the reversal probability per unit time for right- and left-moving particles, while b^+ and b^- give the lowest order approximation to the corresponding densities (number per unit length). In terms of the definitions (39), (38a) and (38b) can be written as

$$\partial b^+ / \partial t = -v \partial b^+ / \partial x + b^- \sigma^- - b^+ \sigma^+, \tag{40a}$$

$$\partial b^- / \partial t = v \partial b^- / \partial x - b^- \sigma^- + b^+ \sigma^+. \tag{40b}$$

Both these equations can be put into the form of the general balance law (3) by making the definitions

$$J^{\pm} = \pm v b^{\pm}, \quad Q^{\pm} = \pm (b^- \sigma^- - b^+ \sigma^+),$$

so that

$$\partial b^{\pm} / \partial t = -\partial J^{\pm} / \partial x + Q^{\pm}.$$

Indeed, J gives the flux of the two species of particles: particles pass a given point at a rate given by the product of their number per unit volume and their velocity, with flux to the right taken as positive. Moreover, Q is truly a net creation term. The number of right-moving bacteria, for example, is decreased by "deaths" when such bacteria reverse and is increased by "births" when left-moving bacteria start to move to the right.

To proceed, we define the overall density (the first approximation to the number of bacteria per unit length) and the net flux J (the first approximation to the net number of bacteria that pass a given point per unit time, with the right-moving bacteria being counted positive):

$$b \equiv b^+ + b^-, \quad J \equiv J^+ + J^-. \tag{41a, b}$$

If we add (40a) and (40b) we obtain

$$\partial b / \partial t = -\partial J / \partial x. \tag{42}$$

Appropriately, this is the *conservation law* for bacteria—left- and right-moving bacterial densities are not separately conserved but we have introduced no mechanism for net bacterial creation. If we subtract (40b) from (40a) we obtain

$$\frac{\partial J}{\partial t} + (\sigma^+ + \sigma^-)J = v^2 \frac{\partial b}{\partial x} + vb(\sigma^- - \sigma^+); \qquad (43)$$

this equation was obtained by Keller (1975), using rather different arguments.

Assumptions that lead to a constitutive law. Later we shall justify the neglect of $\partial J / \partial t$ in most circumstances. If this term is omitted, (43) provides an explicit expression for the flux $J(x, t)$—given the bacterial density b and the reversal probabilities σ at a point x and time t:

$$J = -\left[\frac{v^2}{\sigma^+ + \sigma^-} \right] \frac{\partial b}{\partial x} + \left[\frac{v(\sigma^- - \sigma^+)}{\sigma^- + \sigma^+} \right] b. \qquad (44)$$

Let us now assume that the reversal probability of a given bacterium at a given time depends only on the attractant concentration c and its first temporal derivative at that time:

$$\sigma = f(c, dc/dt). \qquad (45)$$

To justify this assumption, we note that if there were no dependence on any derivatives there would be no history effect. On the other hand, inclusion of higher derivatives can be omitted to give a "simplest" theory in much the same way that higher spatial derivatives were omitted in the postulate (5). Such omissions should be justifiable in the case of sufficiently slow changes.

The bacteria move with speed $\pm v$, so (45) gives

$$\sigma^+(x) = f[c(x), vc'(x)], \quad \sigma^-(x) = f[c(x), -vc'(x)], ' \equiv \frac{d}{dx}. \qquad (46)$$

Again, for sufficiently small gradients it is reasonable to assume that the effect of dc/dt is to cause a relatively small perturbation in the tumbling rate from its basal value at the given concentration $f[c(x), 0]$. This implies that

$$\sigma^{\pm}(x) \approx f[c(x), 0] \pm vc'(x)f_2[c(x), 0], \qquad (47)$$

where f_2 denotes the partial derivative of f with respect to its second argument. Inserting (47) into (44) we obtain a constitutive equation of the form

$$J = -\mu \frac{\partial b}{\partial x} + \chi b \frac{\partial s}{\partial x} \qquad (48)$$

where

$$\mu = v^2/2f, \quad \chi = v^2(-f_2)/f. \qquad (49a, b)$$

For positive chemotaxis we must have $f_2[c, 0] < 0$. But this is just what is observed in the experiments of Berg & Brown (1972). These authors found that (in a spatially homogeneous situation) a rapid temporal increase in chemical concentration brings about much smoother swimming (i.e., decreased tumbling probability) while a decrease in attractant concentration brings about an effect of the opposite character.

A temporal decrease in attractant concentration is observed to have a much weaker influence than a corresponding increase. This can be modelled by assuming that such a decrease has no influence at all, so that

$$\begin{aligned}
\text{if } c' > 0; \quad & \sigma^- = f, \quad \sigma^+ = f + vc'f_2; \\
\text{if } c' < 0; \quad & \sigma^+ = f, \quad \sigma^- = f - vc'f_2.
\end{aligned} \qquad (50)$$

Replacing (47) by (50) has very little effect, however. It only brings about the quantitative change of introducing a factor of $1/2$ into the expression for χ in (49b). Combining (42) and (48) we obtain

$$\frac{\partial b}{\partial t} = \frac{\partial}{\partial x} \left(\mu \frac{\partial b}{\partial x} - \chi b \frac{\partial s}{\partial x} \right). \qquad (51)$$

This equation for the change of bacterial density b is, except for notation, identical to equation (10) for the change of amoeba density a. It should not be surprising that the derivation just given of (51) requires the assumption of sufficiently small attractant gradients. This assumption guarantees features implicitly assumed in the phenomenological derivation of (10) and (51)—features such as lack of influence of higher spatial derivatives and linear dependence on first derivatives. The basic partial differential equation (51) has been used by several authors to study a type of nonlinear travelling wave motion that bacteria can execute under certain conditions. A recent reference is Keller and Odell (1976). Also see Segel and Jackson (1973) and Nossal and Weiss (1973) for the use of this equation in predicting the functional form of μ and χ from experimental data. This last work has been extended by Lapidus and Schiller (1975).

History and inertia effects. Our recent discussion has assumed that the $\partial J / \partial t$ term in (43) is negligible. Let us return to this matter, to see the nature of the effects that this term can bring about and to see when these effects can indeed be neglected.

The presence of the $\partial J / \partial t$ term in (43) means that direct expression for the flux J in terms of other variables is not possible. The fact that J satisfies a differential equation can be ascribed to a sort of inertia. To see this, consider the equation for an undamped spring-mass system:

$$m \frac{d^2r}{dt^2} + kr = f. \tag{52}$$

In (52), if the mass m, and hence the inertia of the system, is very small then the displacement r can be regarded as proportional to the forcing f. Generally, however, the inertia of the system— as expressed through the second derivative term—makes it impossible for the system to respond instantly to its "environment" f. The derivative term $\partial J / \partial t$ in (43) has a similar effect.

Another way of looking at the situation is to regard (43) as a first order differential equation for J. If we assume for simplicity that σ^+ and σ^- are independent of t, we note that (52) has the solution

$$J = \int_{t_0}^{t} \exp\{-(\sigma^+ + \sigma^-)(t - s)\}\left[-v^2 \frac{\partial b}{\partial x} + vb(\sigma^- - \sigma^+)\right]ds$$
$$+ J_0(x, t_0)\exp\{-(\sigma^+ + \sigma^-)t\},$$

where $J_0(x, t_0)$ is the flux at some initial time t_0. Now we have an explicit expression for J, but it involves an integral from the initial time t_0 to the present time t and so it depends on the entire history of the process.* History dependence and inertia are thus two different concepts that can be used in analyzing why the flux generally is not completely characterized by present conditions.

When is the history-inertia term $\partial J / \partial t$ negligible?[†] To answer this question, let us employ a bar to denote the constant scales of the various quantities (Lin and Segel 1974, Section 6.3). Then (42) implies that $\bar{b}/\bar{t} = \bar{J}/\bar{x}$. Using this, we see that the ratio of $|\partial J / \partial t|$ to $|v^2 \partial b / \partial x|$ has the magnitude $[(\bar{J}\bar{b}^{-1})/v]^2$. Thus the first term can be neglected, and inertia effects are not important, when the drift velocity $\bar{J}/\bar{b}$ is small compared to the particle velocity v. This will occur in most cases of bacterial chemotaxis, for the many random tumbles of a bacterium insure that the macroscopic drift of the population is slow compared to the speed of an individual bacterium. In the experiments of Dahlquist, Lovely, and Koshland (1972), for example, the data indicate that J/b is typically about $1/10$ of v.

Some special cases. The most extreme case of inertia is when the transition probabilites σ^+ and σ^- vanish. Then (42) and (43) combine to give the wave equation

$$\frac{\partial^2 b}{\partial t^2} = v^2 \frac{\partial^2 b}{\partial x^2}$$

---

*Note that the form of the kernel $\exp[-(\sigma^+ + \sigma^-)(t - s)]$ ensures a "fading memory" wherein only history for a past interval of time of order $(\sigma^+ + \sigma^-)^{-1}$ is of major significance.

[†]A recent analysis of the possible influence of terms like $\partial J / \partial t$ is provided by Othmer's (1975) discussion of systems of reacting chemicals.

whose solutions never forget their initial shape as they propagate unchanged with speed v.

Another interesting special case is that of the unbiased random walk in which left- and right-transition probabilities are equal. Referring to (39) let us take

$$q^+ = q^- = q \quad \text{so that } \sigma^+ = \sigma^- = q/\tau.$$

By (44) the gross motion is, as expected, one of pure diffusion. The diffusion constant is given by

$$\mu = v^2/(2q). \tag{53}$$

To interpret (53), it is necessary to calculate the mean free path L, i.e., the expected distance between turns. To achieve a run of length $n\Delta$, a particle must persist in the direction it was going for $n - 1$ steps and then reverse; this it will do with probability $p^{n-1}q$. The expected run length is thus given by

$$L = \Delta \sum_{n=1}^{\infty} np^{n-1}q.$$

The standard method of summing this series proceeds by differentiating $G(x)$, the geometric series. We have, for $|x| < 1$,

$$G(x) = \sum_{n=0}^{\infty} x^n = \frac{1}{1-x}, \quad G'(x) = \sum_{n=1}^{\infty} nx^{n-1} = \frac{1}{(1-x)^2}.$$

Thus

$$L = \Delta q G'(p) = \frac{\Delta q}{(1-p)^2} = \frac{\Delta}{q}. \tag{54}$$

Let us denote by T the length of time it takes for the particle to move a distance L without reversing. That is, T is the expected time between turns:

$$T = L/v. \tag{55}$$

From (53), (54), and (55) we have

$$\mu = L^2 / (2T).$$

This is the standard result for the effective diffusion constant in which steps of length L are taken every T seconds, with equal probability to the left or to the right (Lin and Segel, 1974, p. 95). An instance where this special case is relevant is the "constant tumbling" mutants discovered by Spudich and Koshland (1975). These mutants hardly move between tumbles so that L, and hence μ, is very small. Thus "constant tumbling" mutants scarcely move from their initial position—and just this fact was used to detect them.

The previous special case provided a path to a better understanding of the expression for the effective diffusion constant μ. A final case permits similar insight into the chemotactic coefficient χ. Thus, consider a steady-state situation in which bacterial density b is constant, so that (44) reduces to

$$J = \frac{v(\sigma^- - \sigma^+)}{\sigma^- + \sigma^+} b. \tag{56}$$

To understand this result, note that if b^+ and b^- denote the density of right- and left-moving bacteria, we must have

$$b^+ \sigma^+ = b^- \sigma^-, \quad b^+ + b^- = b.$$

(The first of the above equations embodies the fact that in a steady-state situation, the number of right-moving bacteria that reverse must equal the number of left-moving bacteria that reverse.) We now easily find that

$$b^+ = \frac{\sigma^-}{\sigma^+ + \sigma^-} b, \quad b^- = \frac{\sigma^+}{\sigma^+ + \sigma^-} b. \tag{57}$$

But if bacteria move with speed v, then in time Δt those right- (left-) moving bacteria that are within an interval of length $v\Delta t$ to the left (right) of a given line $x = $ constant will cross that line.

Thus

$$J\Delta t = b^+ v\Delta t - b^- v\Delta t. \tag{58}$$

Combining (57) and (58) we obtain (56).

Incorporating the receptor mechanism. We wish to mention the fact that a deeper investigation into the nature of chemotaxis can spring from the basic equations (40). Assume that the tumbling probability for a bacterium depends upon the number of various configurations of receptor molecules located in the bacterial membrane. Concentrations of these molecules change as a result of interaction with the attractant (concentration c) and with each other. Let $C_i^+(C_i^-)$ denote the total number of molecules per unit volume of the ith chemical on all right (left) moving bacteria. By analogy with (40) we have

$$\frac{\partial C_i^\pm}{\partial t} \pm \frac{\partial}{\partial x}\left(v C_i^\pm\right) = R_i\left(C_1^\pm, C_2^\pm, \ldots, c\right) \pm \sigma^- C_i^- \mp \sigma^+ C_i^+,$$

where R_i denotes the rate of chemical reaction. (In many circumstances it would also be necessary to include the effects of diffusion and of consumption of attractant by the bacteria.) The tumbling law takes the form

$$\sigma^+ = f\left(\frac{C_1^+}{n^+}, \ldots, \frac{C_n^+}{n^+}\right), \quad \sigma^- = f\left(\frac{C_1^-}{n^-}, \ldots, \frac{C_n^-}{n^-}\right).$$

To the descriptive earlier theories, we are now adding details of the receptor-attractant interaction. An examination of the implications of such an approach can be found in Segel (1977).

4. APPLICATIONS OF A DIFFERENT MODEL OF RANDOM WALK

Another type of random walk. It is interesting in various biological contexts to consider a random walk in which there is a certain probability $\lambda(x, t)$ of moving *from* the point x at time t, with equal probability that the move is to the right or to the left. Such a situation might apply to an organism that is attracted or repelled to varying degrees by the situation it finds at point x. Let $n(x, t)$ be the number of organisms located at x at time t, and suppose that steps of length Δ are considered every τ time units. Then

$$n(x, t + \tau) = \tfrac{1}{2} n(x - \Delta, t)\lambda(x - \Delta, t) + n(x, t)\left[1 - \lambda(x, t) \right]$$
$$+ \tfrac{1}{2} n(x + \Delta, t)\lambda(x + \Delta, t).$$

By the limiting process (30), we obtain for the population density $a \equiv n/\Delta$ the approximating differential equation

$$\frac{\partial a}{\partial t} = D \frac{\partial^2}{\partial x^2} (\lambda a), \quad D \equiv \frac{\Delta^2}{2\tau}. \qquad (59a, b)$$

This equation is identical in form to the chemotaxis equation (10) if we make the identification

$$\mu = D\lambda, \quad \chi = -D, \quad p = \lambda.$$

Thus the motion described by (59a) is that which would occur if an organism were immersed in a chemical of concentration λ, provided that its diffusivity was proportional to λ by a factor D, and that it was negatively chemotactic to λ with a constant chemotactic factor D.

Skellam (1973) discussed (59a) and related equations in the course of a stimulating article on diffusion processes in population biology. The author and S. Levin (unpublished) have used (59a) to describe "search image" behavior wherein a predator tends to consume prey of the same or nearly the same aspect x as it has been consuming in the recent past.

5. CONCLUDING REMARKS

The various models presented here form a hierarchy exhibiting an increased attention to detail. Slime mold aggregation is approached on a population level. Our major treatment of bacterial chemotaxis is based on a random walk model for the behavior of a single organism. In addition, we have just mentioned an attack on this problem that takes into account the average behavior of the various molecules of attractant located on the bacterial cell wall.

In spite of an increasing involvement with fine structure, each model ultimately yields a set of partial differential equations that purportedly governs overall behavior. Details of individual characteristics are submerged in all cases. One thing, however, that distinguishes biological from physical science is a more pronounced attention to anomaly and uniqueness. A single twitch in a DNA molecule can have fateful consequences for an entire organism—a single cell can be observed for hours in a microscope. In this spirit, Dr. Hanna Parnas and I have recently integrated observations and suggestions of various workers and modelled slime mold amoebae as a collection of points that are capable of moving chemotactically. Each point secretes a pulse of attractant spontaneously and randomly, or when it is subject to a super-threshold concentration of attractant, providing it is not in a refractory period owing to an earlier stimulus. Variability, another factor of profound importance in biology, enters by virtue of a random modulation of all key features of the algorithm that describes a cell's behavior. With the aid of a computer simulation we can watch the behavior of a small collection of cells and can alter our "experimental organism" in accordance with what we see, in order to obtain a better idea of the major factors responsible for a particular kind of behavior. Results are reported in Parnas and Segel (1977, 1978). A general conclusion is that computer simulation can be a valuable tool in contending with biological complexity, provided that considerable information about the system has already been revealed by experiment.

REFERENCES

1. H. C. Berg and D. A. Brown, "Chemotaxis in Escherichia coli analyzed by three-dimensional tracking," *Nature*, **239** (1972), 500–504. Also see an Addendum to a reprint of this article in *Chemotaxis, Biology, and Biochemistry* (E. Sorkin, ed.), S. Karger, Basel, 1975.

2. J. T. Bonner, *The Cellular Slime Molds*, Princeton University Press, Princeton, 1967.

3. S. Corrsin, "Limitations of gradient transport models in random walks and in turbulence," *Advances in Geophysics 18A*, Academic Press, New York, 1974, 25–60.

4. F. W. Dahlquist, P. Lovely, and D. E. Koshland, "Quantitative analysis of bacterial migration in chemotaxis," *Nature New Biol.*, **236** (1972), 120–123.

5. E. F. Keller, "Mathematical aspects of bacterial chemotaxis," *Chemotaxis, Biology, and Biochemistry* (E. Sorkin, ed.), S. Karger, Basel, 1975.

6. E. F. Keller and L. A. Segel, "Initiation of slime mold aggregation viewed as an instability," *J. Theoret. Biol.*, **26** (1970), 399–415.

7. ———, "Model for chemotaxis," *J. Theoret. Biol.*, **30** (1971), 225–234.

8. D. E. Koshland, "Chemotaxis as a model for sensory systems," *FEBS Letters*, **40** (1974), Supplement S3–S9.

9. I. R. Lapidus and R. Schiller, "Bacterial chemotaxis in a fixed attractant gradient," *J. Theoret. Biol.*, **53** (1975), 215–222.

10. C. C. Lin and L. A. Segel, *Mathematics Applied to Deterministic Problems in the Natural Sciences*, Macmillan, New York, 1974.

11. V. Nanjundiah, "Chemotaxis, signal relaying, and aggregation morphology," *J. Theoret. Biol.*, **42** (1973), 63–105.

12. R. Nossal and G. Weiss, "Analysis of a densitometry assay for bacterial chemotaxis," *J. Theoret. Biol.*, **41** (1973), 143–147.

13. ———, "A descriptive theory of cell migration on surfaces," *J. Theoret. Biol.*, **47** (1974), 103–114.

14. G. M. Odell and E. F. Keller, "Necessary and sufficient conditions for travelling bands of chemotactic bacteria," *Math. Biosci.*, **27** (1976), 309–317.

15. H. G. Othmer, "On the significance of finite propagation speeds in multicomponent reacting systems," *J. Chem. Physics*, **64** (1976), 460–470.

16. H. Parnas and L. A. Segel, "Computer evidence concerning the chemotactic signal in aggregating Dictyostelium discoideum," *J. Cell Science*, **25** (1977), 191–204.

17. ———, "A computer simulation of pulsatile aggregation in Dictyostelium discoideum," *J. Theoret. Biol.* (1978), in press.

18. L. A. Segel, "On collective motions of chemotactic cells," *Some Mathematical Questions in Biology*, vol. III (J. D. Cowan, ed.), Amer. Math. Soc., Providence, R.I., 1971, 3–46.

19. ———, "A theoretical study of receptor mechanisms in bacterial chemotaxis," *SIAM J. Appl. Math.*, **32** (1977), 653–665.

20. L. A. Segel, I. Chet, and Y. Hennis, "A simple quantitative assay for bacterial motility," *J. General Microbiol.*, **98** (1977), 329–337.

Lee A. Segel

21. L. A. Segel and J. L. Jackson, "Theoretical analysis of chemotactic movement in bacteria," *J. Mechanochemistry and Cell Motility*, **2** (1973), 25–34.

22. L. A. Segel and B. Stoeckly, "Instability of a layer of chemotactic cells, attractant, and degrading enzyme," *J. Theoret. Biol.*, **37** (1972), 561–585.

23. J. G. Skellam, "The formulation and interpretation of mathematical models of diffusionary processes in population biology," *The Mathematical Theory of the Dynamics of Biological Populations* (M. S. Bartlett and R. W. Hiorns, eds), Academic Press, London, 1973, 63–85.

24. J. L. Spudich and D. E. Koshland, "Quantitation of the sensory response in bacterial chemotaxis," Proc. Nat. Acad. Sci., U.S.A., **72** (1975), 710–713.

25. D. W. Stroock, "Some stochastic processes which arise from a model of the motion of a bacterium," *Z. Wahrscheinlichkeitstheorie und Verw. Gebiete*, **28** (1974), 305–315.

REACTION-DIFFUSION EQUATIONS AND PATTERN FORMATION

N. Kopell

In a pioneering paper in 1952 [1], Turing suggests that some patterns that occur in biology result from an interaction between a chemical reaction and diffusion. Turing concentrates on linear reaction-diffusion equations and shows that they are capable of solutions that are non-homogeneous in space, and which do not decay in time. Since then, there has been a great deal of study of reaction-diffusion equations [2–18], much of it directed toward the existence of non-homogeneous solutions, i.e., pattern. In this paper, I shall be concerned with particular chemical and biological systems in which there are patterns that can be accounted for mainly by reaction and diffusion. These examples show that reaction-diffusion equations can also lead to pattern formation for reasons essentially different from the ones proposed by Turing. In these examples, the nonlinearities in the reaction equations are crucial to the behavior of the solutions.

Reaction-diffusion equations have the form

$$X_t = F(X) + K \, \nabla^2 X \tag{1}$$

where $X = (x_1, \ldots, x_n)$ is a vector of chemical concentrations;

$X = X(\bar{z}, t)$, where t is time and the vector $\bar{z}$ is the spatial variable (which may have any number of dimensions). $X_t = F(X)$ are the kinetic equations which describe the course of the chemical reaction. K is a positive definite matrix of diffusion coefficients, and the term $K \nabla^2 X$ is the standard model for diffusion (as in the heat equation).

In [1] Turing considers a case of equation (1) in which $\bar{z}$ is one dimensional and periodic (so his space is a circle). His kinetic equations are linear; that is, they have the form

$$X_t = M(X - X_0)$$

where M is a matrix. (For convenience we may translate the origin so the critical point X_0 is at $X = 0$.) The critical point is assumed to be asymptotically stable; i.e., all the eigenvalues of M have negative real part, so that, without diffusion, all disturbances decay. It might seem that diffusion could only help to smooth out any disturbance and return the system to its uniform state $X \equiv 0$. However, as Turing shows, if the diffusion matrix K is far from being scalar (so that some of the chemicals diffuse much faster than others), diffusion may actually help some disturbances to grow. A simple related example of this occurs in a discrete version of equation (1) with two adjoining cells exchanging material across a membrane. If we label the concentrations in cell i by X^i then the appropriate equations are

$$\begin{aligned}
X_t^1 &= MX^1 + K(X^2 - X^1), \\
X_t^2 &= MX^2 + K(X^1 - X^2).
\end{aligned} \tag{2}$$

Even if the eigenvalues of the $n \times n$ matrix M all have negative real parts, the $2n \times 2n$ matrix

$$\begin{pmatrix} M - K & K \\ K & M - K \end{pmatrix}$$

of the system (2) may have some eigenvalues with positive real part; thus disturbances from the equilibrium concentration $X^1 =$

$X^2 = 0$ will grow. (Turing cited the 2×2 example

$$M = \begin{pmatrix} 5 & -6 \\ 6 & -7 \end{pmatrix} \qquad K = \begin{pmatrix} .5 & 0 \\ 0 & 4.5 \end{pmatrix}.$$

A non-linear version of equations (2) studied by S. Smale [16] has an oscillatory solution.)

For a ring of discrete cells or continuous tissue, Turing shows that equation (1), with F linear, can have solutions which are wavelike in character. Essentially, he writes the solutions as Fourier series

$$X = \sum_{n = -\infty}^{\infty} A_n e^{inz} e^{p_n t},$$

and focuses on the term whose time eigenvalue p_n has the largest real part; the amplitude of this term grows the fastest and so it might be expected to dominate. If p_n is real

$$A_n e^{inz + p_n t} \tag{3}$$

can be thought of as a spatially periodic concentration variation in the circle of tissue, whose amplitude grows in time. If p_n is complex with a positive real part, (3) is a traveling wave with a growing amplitude. The rationale of the analysis is that, although these solutions are not stable (the amplitudes grow without bound), the behavior at the onset of instability may give a clue to the later pattern. (Of course the non-linearity must play a role in the final pattern.)

In Turing's model, the instability is due basically to a difference in diffusion coefficients, which encourages certain kinds of disturbances to grow. (For a more detailed exposition of this phenomenon, see [15].) Similar phenomena occur even if the spatial variable is in a Euclidean space instead of a circle [13]. (In the context of more complicated hydrodynamic equations, this "double diffusion" phenomenon is very familiar [19] if not quite fully understood.)

A class of biological models related to Turing's ideas are those of Gierer and Meinhardt [5, 6] for pattern formation based on lateral inhibition. The "patterns" they study are not primarily waves, but the development of stable, inhomogeneous distributions (gradients or isolated peaks) of substances. For example, the retina of the amphibian Xenopus develops a definite polarity along two perpendicular axes. Jacobson and Hunt showed that if the retina is rotated by 180° after some critical time in development (different for each axis), it retains its polarity in spite of its environment; if rotated before this stage, it changes its polarity to match that of the host tissue. Furthermore, if the retina is removed before time t_0 and developed in tissue culture, and then transplanted after time t_0 in reverse orientation, the retina retains its polarity, just as if it had stayed in the original environment through time t_0. (For references, see [6].)

These experimental observations are modelled by Gierer and Meinhardt using a pair of diffusing and reacting substances, an "activator" with a lower diffusion constant, and an "inhibitor" with a much higher diffusion constant; the polarity of the tissue is represented by the gradients that form in these substances. Their equations are somewhat more complicated than (1) since the reaction rates depend on spatial position. One version of their equations is as follows:

$$a_t = \rho + c\rho a^2/h - \mu a + D_a a_{zz},$$
$$h_t = \rho' a^2 - \nu h + D_h h_{zz} + \tilde{\rho}(z). \tag{4}$$

Here z is one dimensional (each spatial axis is considered separately); a and h are the concentrations of the activator and inhibitor; ρ and ρ' are measures of the sources of activator and inhibitor. (They are constant for this example, in which the tissue has no intrinsic polarity.) The term $c\rho a^2/h$ represents the autocatalytic growth of the activator concentration, which is reduced by the presence of the inhibitor; the inhibitor grows at a rate proportional to a^2. The terms $-\mu a$ and $-\nu h$ represent the degradation of the chemicals. The term $\tilde{\rho}$ depends on the spatial position z and models the influence of the surrounding tissue; this term is linear in z with a very small slope. (In other versions of these equations, $\tilde{\rho} = 0$ and ρ and/or ρ' are functions of z.)

The above equations have so far been studied mainly numerically, and we do not have a complete picture of their behavior. Nevertheless, the extensive calculations made by Meinhardt and Gierer are strong support for a model such as (4). An interesting feature of the model is that the time at which the polarity is determined appears to be relatively independent of the details of the original asymmetry; thus the model provides a kind of time regulation. A variant of the model is capable of size regulation in the following sense: if a section of a final pattern, with a small gradient, is removed and allowed to evolve, it develops a steeper gradient with the same polarity and qualitative shape as the original pattern.

The difference in the diffusion coefficients seems to play the same role in these complicated equations as in the simpler Turing equations. More specifically, for an appropriate set of parameters, there is a unique steady state for the kinetic equations associated with (4) (with $\tilde{\rho}$ constant), and this steady state is stable to homogeneous perturbations. (Some parameters mentioned by Gierer and Meinhardt, after scaling, are: $\rho = 5.625$, $c\rho = 1$, $\mu = 56.25$, $D_a = 1$, $\rho' = 1$, $\nu = 78.75$, $D_h = 22.5$, $\tilde{\rho} = .25$.) It is easy to show that if $D_h/D_a \equiv L$ is close to one, the homogeneous solution of (1) (with $\tilde{\rho}$ constant) is stable as a solution to the boundary value problem $a_z = 0 = h_z$ at $z = 0$ and $z = 1$. But if L is sufficiently large (greater than $L_0 \sim 11.2$ for the figures cited above), this solution is unstable. Bifurcation techniques [20] show that there are then other stable stationary states; for L not much greater than L_0, the steady solutions $a(z)$ and $h(z)$ are approximately of the form constant + constant $\cdot \cos \pi z$. Due to the symmetry in the problem there are two such solutions.

Now if $\tilde{\rho}$ is not constant, or if ρ or ρ' is a function of z, we have a "perturbed bifurcation problem". If L is somewhat below L_0, the "almost constant" stationary state is stable. As L increases, there is a continuation of this almost constant state which is close to one of the stable stationary solutions to the homogeneous problem. Above some value of L near L_0, there are two more stationary states, one of which is stable (and near the other stable stationary state of the homogeneous problem). For homogeneous initial conditions, the evolution is biased toward the continuation of the almost constant state. But if the initial conditions are themselves sufficiently nonconstant, the other stable steady state may be chosen. This helps to explain the simulation of the transplantation data mentioned above: once a gradient in a or h is sufficiently big, if their reversal is taken as initial conditions, the small opposing gradient in $\tilde{\rho}$ will have no effect on the qualitative outcome. (See Figure 1.)

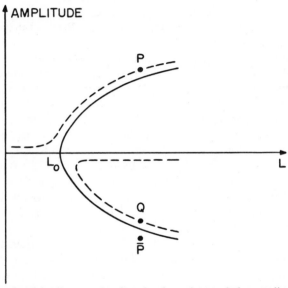

FIG. 1. A schematic diagram showing the dependence of the amplitude of the steady state solution on the ratio of the diffusion coefficients (near the bifurcation value). The amplitude may be measured by $\int_0^1 [a(z) - a_0]\cos \pi z \, dz$, where $a \equiv a_0$ is the uniform solution. The solid curve represents the amplitude of the two nonuniform stationary steady states for the unperturbed bifurcation problem. The dotted line shows the amplitudes of the nearly steady states if any of the coefficients are functions of z. (The picture is drawn for the case in which the gradients in the source terms bias the system toward the steady state with positive amplitude.) Point P represents the amplitude of a steady solution $(a(z), h(z))$, and point $\bar{P}$ that of the initial conditions $(a_0 + [a_0 - a(z)], h_0 + [h_0 - h(z)])$. With these initial conditions, the system will tend to a steady state Q which has a gradient in the same direction as the initial conditions, in spite of the opposite bias of the system.

Gierer and Meinhardt have studied many models related to (4). The form of the kinetic equations may be varied to fit different biochemical assumptions (for example that the activator and inhibitor are related by chemical conversion). The equations can also be refined to include growth and differentiation on a slower time scale; Meinhardt and Gierer use this to model head formation in hydra.

The numerical experiments indicate that equation (4) and related equations can exhibit many different kinds of qualitative behavior. For example, if the inhibitor equilibrates slowly (i.e., ν is small), there may be oscillations. If growth is included (by adding on to the spatial interval as time increases), equation (4) (with $\tilde{\rho} = 0$) is capable of steady state solutions which have spatially periodic peaks of activator concentration. (The rationalization is that, as the size of the interval increases, it exceeds the range of the inhibitor, and another region of high activator concentration can occur.) These and other phenomena of those equations can be partially attacked by bifurcation methods and raise many interesting mathematical questions which have yet to be fully explored. (For related questions, see [4] and [9].)

A pattern formation problem whose mathematics has been much more studied is not biological, but chemical. It involves an oscillating chemical reagent which was discovered by Belousov [21] in 1958. The main reaction is the bromination of malonic acid, which is catalyzed by cerium in an oscillatory manner; if an appropriate oxidation reduction indicator is added, the system turns alternately red and blue with a period of about a minute (depending on initial concentrations).

This reagent is capable of producing several different kinds of patterns. If the fluid is in a vertical tube, with a gradient in temperature or in one of the initial concentrations, horizontal bands form and propagate through the tube. These bands do not rely on diffusion as the mechanism for pattern formation; the bands can be explained on a purely kinematic basis [22]. If the fluid is placed in a thin layer and heated from below or cooled from above (for example, by evaporation), cells appear in the fluid which are reminiscent of the Benard cells which form in similar hydrodynamic circumstances. These cells are still somewhat mysterious; even if the fluid is convecting (in Benard cells), it is not clear what coupling between the hydrodynamics and the chemistry causes the boundaries of the cells to be so clearly outlined in dark red.

The Belousov reagent also produces another set of patterns first reported by Zaiken and Zhabotinsky [23]. These patterns form when the reagent is placed in a thin layer and covered to prevent convection. Blue spots appear and the blue region propagates outwards; then the center turns red and this region propagates outward. This process repeats itself over and over, and after five or ten minutes, the entire surface is covered with patterns of concentric rings ("target patterns"). Winfree [24] showed that the fluid can be modified so that it no longer oscillates spontaneously, but is still capable of propagating waves, e.g., if it is "poked" with a hot wire. I shall refer to the modified reagent (which Winfree calls the Z reagent) as excitable, and the original reagent as oscillatory. Both reagents form spirals as well as concentric ring patterns. (Spirals can be obtained from the target patterns by perturbing the fluid a little.)

Although some features of the overall pattern are quite variable (such as the number and position of the centers), there are other features which are always present. For example, in each such target pattern, the frequency of oscillation at each point in the fluid, as well as the spacing and propagation speed of the bands, is constant. (All three quantities change from target pattern to target pattern.) Where the target patterns meet each other (or where the rings propagate into the spatially homogeneous fluid) there is a "shock front" across which there is an abrupt transition in frequency and wave length.

For the target patterns and spirals, there is widespread agreement that the mechanism behind the formation involves just reaction and diffusion. The above observations raise the mathematical question: what properties must the kinetic equations and the diffusion matrix have in order for there to be solutions to (1) which behave as above? This question can be made sharper by noting that different portions of the fluid appear to develop independently of other regions, with interactions only in thin "boundary layers". (In particular, the boundaries of the container appear to have no effect a short distance away.) Because of this, we hypothesize that, if the evolution of the patterns is governed by equations of the form (1), there should be solutions to (1) representing idealizations of various portions of the fluid.

For example, in a neighborhood of a radial line in one of the target patterns, away from the center and boundary of the set of concentric rings, the pattern is that of a plane wave; thus a plane wave solution is a simple idealization of a portion of the overall pattern. Other, more complicated idealizations, are single, infinitely large target patterns (with infinitely many rings), or single, infinitely large spiral patterns. (A one spatial dimension version of the target pattern is a pair of plane waves of the same frequency, heading outward from a central transition region.) Another important idealization involves the interaction between two target patterns; if there are many rings in each target pattern then, near the line joining the centers, the pattern is close to that of two infinite periodic plane waves meeting at a shock front. Note that these more complicated idealizations are all locally like plane waves except at some singular sets: the center of a target pattern or spiral, and the shock fronts between the patterns.

The simplest mathematical problem concerns periodic traveling waves. These correspond to solutions to (1) which have the form

$$X(\bar{z}, t) = Y(\sigma t - \bar{\alpha} \cdot \bar{z}). \tag{5}$$

Here Y is a 2π periodic function of its argument, σ is the frequency, and $\bar{\alpha}$ is the wave number vector. (This is a more restrictive notion of periodic wave than Turing's; (5) has a wave form which does not change in time.) Y satisfies a system of ordinary differential equations of dimension $2n$ if X is a vector of dimension n. This system is

$$\sigma Y' = F(Y) + \alpha^2 K Y'', \tag{6}$$

where $'$ means differentiation with respect to $\sigma t - \bar{\alpha} \cdot \bar{z}$ and $\alpha = |\bar{\alpha}|$.

The most general result [10] about the existence of such solutions to (1) says that if the kinetic equations have a stable limit cycle solution, then (1) has a one-parameter family of traveling wave solutions, parameterized by α. The argument guarantees the solutions only for α small. Examples given in the same paper suggest that the hypothesis that α be small is stronger than necessary.

Of course, the excitable reagent does not have a stable limit cycle. So far, there has been no general definition of "excitable". Various models of kinetic equations have been studied which do not have periodic solutions, but which do have wave solutions to (1). These models all have a stable critical point and the property that some orbits starting near this critical point go far away before returning (as $t \to \infty$).

One of the models [18, 25], studied numerically by Winfree, is suggested by analogy to the Nagumo equations for nerve impulse transmission [26, 27; see these for further references]. The equations are:

$$\dot{x}_1 = -\left(x_1^3 - 3x_1 + x_2\right),$$

$$\dot{x}_2 = \epsilon(x_1 - \lambda). \tag{7}$$

Here ϵ and λ are parameters, with ϵ small, and the parameter λ affects the qualitative nature of the kinetic equations: for $\lambda < 1$, there is a stable limit cycle and so the above result applies. For $\lambda > 1$, the kinetic equations no longer have an oscillatory solution, but (1) still has periodic traveling waves, at least for $\lambda \leqslant \sqrt{3}$. According to G. Carpenter this can be shown by techniques used in [26]. Again, the hypotheses for these techniques are probably somewhat more restrictive than necessary; the methods work only when some of the diffusion coefficients are very small relative to others.

Another model that has been the subject of a great deal of work is the Field–Noyes model [28]. Field and Noyes, along with Körös [29], did the most extensive experiments on the mechanism of the reaction. The mechanism is extremely complicated, but Field and Noyes think its major features are modelled by the following three-dimensional system:

$$\dot{x}_1 = s\left(x_1 + x_2 - x_1 x_2 - q x_1^2\right),$$
$$\dot{x}_2 = s^{-1}\left(-x_2 - x_1 x_2 + f x_3\right), \qquad (8)$$
$$\dot{x}_3 = w\left(x_1 - x_3\right).$$

For some range of the parameters s, q, f, and w, this system has a stable limit cycle; for others it does not. Oscillatory solutions to the kinetic equations (8) have been investigated by Noyes and Field [28], Stanshine [17], and Hastings and Murray [30]; plane wave solutions to the associated equations (1) were found by Stanshine [17], in cases where (8) has oscillatory solutions and in other cases in which it does not.

Closely related to periodic plane wave solutions are isolated pulse solutions. These are solutions to (1) which have the form

$$X(\bar{z}, t) = Y_1(\bar{z} - ct),$$

where Y_1 tends to the same constant for fixed $\bar{z}$ as $t \to \pm \infty$. Y_1 satisfies essentially the same ordinary differential equation (6) as the periodic waves, but with different conditions at $t = \pm \infty$. Pulse solutions have been found for various models of the kinetics [17, 26, 31, 32]. They are limits of periodic traveling waves, for α and σ tending to zero, but $\sigma/\alpha = c$ staying finite. For fixed kinetic equations there appears to be only a finite number (generally one or two) of such pulses; for the oscillatory reagent there may be none [10]. It is a reasonable conjecture that, as in the Nagumo equations, only one of these pulses is stable as a solution to equation (1). But so far, questions about stability of the waves and pulses remain open. (Some results in that direction are in [10] and [17].)

The actual patterns observed in the Belousov reagent are not plane waves, but they are locally like periodic plane waves. Thus, whatever hypotheses on the kinetic equations are needed to produce the traveling waves are also necessary for the more complicated patterns. In this spirit, Greenberg [7] showed that if there are plane wave solutions to (1) for all α sufficiently small (equivalently, if the kinetic equations have a limit cycle), then there are solutions to (1) which behave asymptotically (for large $|\bar{z}|$) like an infinite spiral. (His solutions are not defined near $\bar{z} = 0$.) However, the existence of the plane waves is not sufficient to insure that of the more complicated solutions, in particular the solutions representing waves meeting along a shock front ("shock structure" solutions). This problem has been examined [12] under the hypothesis that there are plane wave solutions to equation (1) (e.g., when the kinetic equations have a limit cycle). It was found that the existence of shock structure solutions appears to impose a restriction on the dispersion function $\sigma = H(\alpha^2)$ relating the frequency and wave number in the one-parameter family of traveling wave solutions; this restriction is analogous to the entropy condition for shock structures in gas dynamics (which rules out expansion shocks), and places further conditions on the nonlinear terms in the kinetic equations. (For more details, see [11] or [12].)

The question of whether there are solutions to (1) which represent a single (infinite) target pattern or spiral, and which are regular at the center, has been less examined. Recent work of the author and L. N. Howard strongly suggests that if the kinetic equations have the non-linearity needed to produce the shock solutions, then the reaction-diffusion equations do not have a one parameter family of target pattern or spiral solutions (i.e., one for each α or σ in some range). (Also see [14].) For simple model equations introduced in [10], the problem (or at least a one dimensional version of it) turns out to be a non-linear eigenvalue problem: only for certain discrete values of σ are there such solutions. Each solution is asymptotically (large $|\bar{z}|$) one of the plane wave solutions to (1). At most finitely many of these target pattern solutions are stable. We conjecture further that when there are impurities at the centers of the patterns, the particles act as a boundary condition for the rest of the pattern which makes possible a different set of values of σ; thus, in the reagent, if there are many heterogeneities, one may see a whole range of frequencies.

In the formation of the Zhabotinsky–Zaiken patterns, the difference in diffusivities which was so crucial to the Turing model plays very little role. (At most, the ratio between two diffusivities may provide a small parameter which is useful in making calculations.) The crucial feature in the pattern formation is that the kinetic equations have a limit cycle or a trajectory which is topologically "almost" a limit cycle. The equations studied by Gierer and Meinhardt do use the differences in diffusion coefficients. But beyond this similarity to the Turing model, there are fundamental contrasts in philosophy. Turing's patterns arise from a homogeneous fluid by instability. The Gierer–Meinhardt approach is to show that highly structured and reproducible pattern can arise because of already existing biases which are amplified and regulated by non-linear terms in the kinetic equations. (For a related approach, in terms of networks, see [34].) These two classes of model problems give some hint of the richness in the mathematical behavior of reaction-diffusion equations.

Acknowledgement: This work was supported by the National Science Foundation under Contract No. GP-335-49A-1 and the M.I.T. Mathematics Department.

REFERENCES

1. A. M. Turing, "The chemical basis of morphogenesis," *Philos. Trans. Roy. Soc. B*, **237** (1952), 37.

2. R. Aris and K. H. Keller, "Asymmetries generated by diffusion and reaction, and their bearing on active transport through membranes," *Proc. Nat. Acad. Sci.*, **69** (1972), 777.

3. J. F. G. Auchmuty and G. Nicolis, "Bifurcation analysis of non-linear reaction-diffusion equations," Preprint.

4. J. Boa, "A model biochemical reaction," Thesis, Cal. Tech., 1974.

5. A. Gierer and H. Meinhardt, "A theory of biological pattern formation," *Kybernetic*, **12** (1972), 30.

6. H. Meinhardt and A. Gierer, "Applications of a theory of biological pattern formation based on lateral inhibition," *J. Cell Sci.*, **15** (1974), 321.

7. J. Greenberg, "Periodic solutions to reaction diffusion equations," *SIAM J. Appl. Math.*, **30** (1976), 199.

8. J. Guckenheimer, "Constant velocity waves in oscillating chemical reactions," in *Structural Stability, The Theory of Catastrophes and Their Application in the Sciences*, Springer Lecture Notes, No. 525.

9. M. Herschkowitz-Kaufman, and G. Nicolis, "Localized spatial structures and non-linear chemical waves in dissipative systems," *J. Chem. Phys.*, **56** (1972), 1890.

10. N. Kopell and L. N. Howard, "Plane wave solutions to reaction-diffusion equations," *Studies in Appl. Math.*, **52** (1973), 291.

11. L. N. Howard and N. Kopell, "Wave trains, shock fronts and transition layers in reaction-diffusion equations," *SIAM-AMS Proc.*, **8** (1974), 1.

12. L. N. Howard and N. Kopell, "Slowly varying waves and shocks in reaction-diffusion equations," *Studies in Appl. Math.*, **56** (1977), 95.

13. H. Othmer and L. E. Scriven, "Interactions of reaction and diffusion in open systems," *Ind. and Eng. Chem. Fund.*, **8** (1969), 302.

14. P. Ortoleva and J. Ross, "On a variety of wave phenomena in chemical reactions," *J. Chem. Phys.*, **60** (1974), 5090.

15. L. Segel and J. Jackson, "Dissipative structure: an explanation and an ecological example," *J. Theor. Biol.*, **37** (1972), 545.

16. S. Smale, "A mathematical model of two cells via Turing's equation," *Lectures on Mathematics in the Life Sciences*, **6** (1974), 17.

17. J. Stanshine, "Asymptotic solutions of the Field-Noyes model for the Belousov reaction," Thesis, M.I.T., 1975.

18. A. T. Winfree, "Rotating reactions," *Scientific American*, **230** (1974), 82.

19. J. S. Turner, *Buoyancy Effects in Fluids*, Cambridge University Press, New York, 1973.

20. D. H. Sattinger, "Topics in Stability and Bifurcation Theory," *Lecture Notes in Mathematics*, No. 309, Springer-Verlag, New York, 1969.

21. B. Belousov, Sb. ref. radiats. med. za. Medzig, Moscow, 1959.

22. N. Kopell and L. N. Howard, "Horizontal bands in the Belousov reaction," *Science*, **180** (1973), 1171.

23. A. N. Zaiken and A. M. Zhabotinsky, "Concentration wave propagation in two-dimensional liquid-phase self-oscillating system," *Nature*, **225** (1970), 535.

24. A. T. Winfree, "Spiral waves of chemical activity," *Science*, **175** (1972), 634.

25. E. C. Zeeman, "Levels of structure in catastrophe theory," Invited Address International Congress of Mathematicians, Vancouver, 1974.

26. G. Carpenter, "Traveling wave solutions to nerve impulse equations," Thesis, Univ. of Wisconsin, 1974.

27. S. P. Hastings, "The existence of periodic solutions to Nagumo's equations," *Quart. J. Math.*, **25** (1974), 369.

28. R. J. Field and R. M. Noyes, "Oscillations in Chemical Systems, IV," *J. Chem. Physics*, **60** (1974), 1877.

29. R. J. Field, E. Körös and R. M. Noyes, "Oscillations in Chemical Systems, I," *J. Am. Chem. Soc.*, **94** (1972), 8649.

30. S. P. Hastings and J. D. Murray, "The existence of oscillatory solutions in the Field-Noyes model for the Belousov-Zhabotinskii reaction," *SIAM J. of Appl. Math*, **28** (1975), 678.

31. N. Kopell and L. N. Howard, "Pattern formation in the Belousov reaction," *Lectures on Mathematics in the Life Sciences*, **7** (1974), 201.

32. S. P. Hastings, "On the existence of homoclinic and periodic orbits for the Fitzhugh- Nagumo's equations," *Quart. J. Math.*, **27** (1976), 123.

33. A. T. Winfree, "Scroll shaped waves of chemical activity in three dimensions," *Science*, **81** (1973), 937.

34. S. Grossberg, "On the development of feature detectors in the visual cortex with applications to learning and reaction-diffusion systems," *Biol. Cybernetics*, **21** (1976), 145.

PRIMARY AND SECONDARY WAVES IN DEVELOPMENTAL BIOLOGY

E. C. Zeeman

ABSTRACT*

Using catastrophe theory, we prove a theorem to the effect that whenever a multicellular mass of tissue differentiates into two types, the frontier between the two types always forms to one side of its final position, and then moves through the tissue before stabilising in its final position. We call this movement a primary wave. Primary waves may sometimes be identified as hidden waves of cell determination, which may not manifest themselves visibly until after a delay of several hours. The visible manifestation will then be a secondary wave of cellular activity, which may cause morphogenesis, for example rolling changes of curvature.

Two applications are worked out in detail, namely models for gastrulation and neurulation of amphibia, and for culmination of cellular slime mold. In the amphibian model the differentiation

*Editor's note: This selection represents a portion of a longer paper which appeared in S. A. Levin, Ed., Lectures on Mathematics in the Life Sciences 8, Some Mathematical Questions in Biology 7, AMS, Providence, Rhode Island, 1974. The original full abstract and contents are retained, but the bibliography is abbreviated.

between ectoderm and mesoderm causes a hidden primary wave, whose visible secondary wave of cells submerging causes not only the morphogenesis of gastrulation but also the formation of notochord and somites. In the slime mold model the differentiation between spore and stalk causes a hidden primary wave, whose visible secondary wave of cells submerging causes culmination and the morphogenesis of the fruiting body. Both models suggest experiments by which they can be tested.

1.   INTRODUCTION

Our objective is to explain primary waves by catastrophe theory [18], and secondary waves by cell physiology [1, 8], and then to use them both together to explain morphogenesis.

By a *wave* we mean the movement of a frontier separating two regions. We call the wave *primary* if the mechanism causing the wave depends upon space and time. We call the wave *secondary* if it depends only upon time, in other words it is series of local events that occur at a fixed time delay after the passage of the primary wave. Therefore, whereas the wave-form of the primary wave is fundamental, the secondary wave only appears to have a wave-form because it follows the primary wave after the fixed time delay. In a sense the wave-form of the secondary wave is accidental because it could be disrupted by mixing up the substrate in between the passage of the two waves. The epidemic example in §2 below illustrates this point. The point is further emphasised by the following difference between the primary and secondary waves: if the substrate is cut before the passage of the primary wave then this stops the primary wave. However if the cut is made between the passage of the two waves then this will not stop the secondary wave, which will appear to jump across the cut.

If the primary wave is invisible, then the secondary wave may appear mysterious. We suggest that this may be a typical situation in developmental biology. For instance a primary wave across a multicellular mass of tissue might consist of the switching on of certain gene systems in each cell, and this may be difficult to detect at the time because biochemical analysis tends to disrupt the delicate dynamics; in fact most experimental evidence that gene systems have been switched on seems to come from observation of some secondary effect after a suitable time delay (see §7 below). The secondary effect in this case is usually some physical manifestation in cell behaviour such as change in chemical composition, change in RNA content, change in oxygen consumption, change in membrane cohesiveness, change in shape, change in amoeboid activity, change in mitosis rhythm, etc. Another common and important secondary effect is for the cell to alter the ratio between the areas of that part of its membrane in contact with other cells and that part comprising free surface of the tissue; for instance the cell can increase contact with other cells by amoeboid action towards them, and decrease its free surface by wrinkling its free membrane (see Figure 21). For convenience we call this process *submerging*. For example, submerging happens during gastrulation (see Figure 20).

It is the secondary wave of physical manifestation that may signal the release of chemical energy to provide the physical energy necessary for morphogenesis. For example submerging cells may push and pull on their neighbours, and thereby alter the overall curvature of the free surface, as described in Gustafson & Wolpert [8]. We suggest that some morphologies that hitherto may have appeared to be explicable may now be explained in terms of secondary waves. If this is the case, this may provide a conceptual framework for the experimental search for hidden primary waves.

The next question is: what causes a primary wave? The simplest mechanism is diffusion, for instance of chemicals or signals. In order to illustrate the difference between primary and secondary waves, we briefly give elementary examples of epidemics and regulation in §§ 2 and 3, in which the primary wave is caused by diffusion.

However for the rest of the paper we are interested in a more subtle mechanism for producing primary waves. We prove a theorem that the four hypotheses

I Homeostasis
II Continuity
III Differentiation
IV Repeatability

the primary wave begins at the tip and proceeds $\frac{1}{3}$ of the way along the grex, several hours before culmination. We then deduce the morphogenesis of the fruiting body, giving predictions as to shape and speed.

I am indebted to many people for discussions, particularly mathematicians, René Thom, David Fowler, and Klaus Jänich, and biologists C. H. Waddington, Jack Cohen, Lewis Wolpert, Peter Nieuwkoop, Jonathan Cooke, and John Ashworth. The main inspiration came from years of conversations with René Thom about applying catastrophe theory to biology. Meanwhile in counterconversations Lewis Wolpert emphasised the inadequacy of using catastrophe theory by itself, because it can only explain the geometry, and not the forces that shape the embryo. On the other hand looking at the local forces by themselves cannot explain the global geometry. Hence the concept of primary and secondary waves grew out of trying to put these two ideas together, the mathematical and the biological.

Discussions with Peter Nieuwkoop about gastrulation were particularly valuable during the germination of the ideas, after an initial presentation of the theorem at a conference in Göttingen in September 1973 organised by Klaus Jänich. Jonathan Cooke stimulated the ideas about pattern-formation and the somites. John Ashworth explained to me the slime mold morphogenesis. I an indebted to the A.A.A.S. and A.M.S. for the opportunity to present the ideas, and to the various authors and journals for permission to reprint their diagrams.

CONTENTS

2. EXAMPLE: EPIDEMIC

This is a simple example to illustrate the difference between primary and secondary waves. The substrate is people. The two regions are those who have been infected by the epidemic and those who have not. The frontier bounds the infected region. This frontier moves forward as a hidden primary wave of infection. This is a simple diffusion wave, that moves steadily forward at the speed of diffusion as each person infects his neighbours (similar to Huyghen's principle). Then after a fixed time delay the visible secondary wave of symptoms follows, assuming that the people have remained stationary. If the people move about in between the waves then the wave-form of the secondary wave will be disrupted. If the substrate is cut before the arrival of the primary wave, in other words the infected population is quarantined, then this stops the primary wave, the spread of infection. However if the substrate is cut between the waves, in other words only the population already showing symptoms is quarantined, then this does not stop the secondary wave, the spread of symptoms.

3. EXAMPLES: REGULATION

In the heartbeat the pacemaker wave causing muscular contraction is a primary wave, and about half a second later the wave muscular relaxation is a secondary wave [24]. In the nerve impulse the membrane depolarisation along an axon is a primary wave, and about a millisecond later the repolarisation is a secondary wave. On both these cases the primary wave is electro-chemical, and could be described as a diffusion wave (diffusion of electrons) proceeding according to Huyghen's principle.

The main feature of a diffusion wave is that it preceeds at constant speed. By contrast the main feature of the more subtle kind of primary wave that the theorem describes, and which we shall be considering from now on, is that it slows to a halt; in other words the frontier stabilises.

4. EXAMPLE: ECOLOGY

This is a simple example to illustrate the more subtle kind of primary wave. The wave has complicated causes, but is easy to understand because it is visible rather than hidden. The ecology is greatly oversimplified, but then we are only using it to illustrate the idea.

Consider the ecological development of grass and trees over a continuous environment of soil and climate. Suppose for simplicity that the northern end of the environment is suitable for grass only and the southern end for trees only, so that the former eventually develops into mature grassland, and the latter into mature forest. Suppose that as either vegetation gets established it suppresses the other; trees fail to survive in grassland and grass fails to survive in forest. Suppose at first there is a continuous variation of vegetation, varying from forest in the south, with trees gradually thinning as we proceed north, until grassland is reached. Then at time t_1 the forest will develop a noticeable frontier at latitude s_1, say. This frontier will deepen, in the sense that the difference between the two sides of the frontier will become more marked, due to the suppressive effect of either vegetation upon the other. As the frontier deepens it would be exceptional for it to remain at s_1, the place where it originally formed (exceptional from the point of view of repeatability, as we explain below). Therefore, depending upon the initial conditions, it will either move north as the mass of trees seed themselves into the grassland,

214   *E. C. Zeeman*

or move south as the grassland erodes the forest edge. Suppose that in our case the initial conditions are such that the frontier moves north as in Figure 1.

Eventually the northerly expansion of the forest balances out against the unsuitability of the northern climate for trees, and so the northerly movement of the frontier slows down, until it stabilises at s_2 at time t_2. Thereafter the frontier remains in stable equilibrium, and deepens further. The movement of the frontier from s_1 to s_2 during the time interval $t_1 < t < t_2$ is the primary wave.

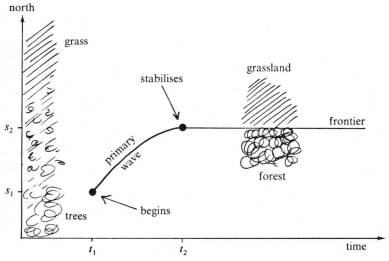

FIG. 1. The frontier of a forest moving as a primary wave.

The primary wave is succeeded by a series of secondary waves representing the spread of various species of flora and fauna that require various time delays of maturity before the forest becomes a suitable habitat for them (woodworm prefer old trees). However in this example we are less interested in the secondary waves.

To show that the primary wave is an illustration of the theorem, we must interpret the four hypotheses in this case.

I:Homeostasis is the tendency of the vegetation in any one place to develop into a stable state, stable with respect to time, that we can name as grassland or forest.

II:Continuity refers to both the continuous environment of soil and climate, and the initial continuous variation of vegetation.

III:Differentiation means that at the end there are two distinct states, mature grassland and mature forest.

IV:Repeatability means that if the initial conditions are varied slightly then the values s_1, t_1, s_2, t_2 may vary slightly but the qualitative behaviour of the frontier remains the same. In other words repeatability means that the whole space-time development, or *chreod* [22], will be stable under sufficiently small perturbations of the initial conditions. Repeatability is an essential hypothesis for the existence of the wave, because it implies $s_1 \neq s_2$. Otherwise, if $s_1 = s_2$, meaning that the frontier had stabilised where it formed, and so causing no wave, then this would be unrepeatable, in the sense that the initial conditions must have been exceptional, and an arbitrarily small perturbation of them could cause $s_1 \neq s_2$, and hence cause a wave, in other words a qualitatively different development.

Notice that in this example of a primary wave, the wave could be said to be caused by diffusion as the trees seed themselves into the grassland. However the situation is not as simple as in the previous examples, where the speed of the wave was constant and equal to the speed of diffusion, because here the wave slows down and stops as the frontier stabilises. One could make an elementary model of this slowing down by using a linear differential equation with a diffusion term balanced against a survival term, but this would not give insight into the formation and the deepening of the frontier, as does the more sophisticated catastrophe model. Also the two models give different quantitative predictions, which would distinguish between them: for instance in the elementary diffusion model the wave slows down exponentially, but in the catastrophe model it slows down parabolically (see §9). The catastrophe model is more likely to be correct, because of the two processes involved, the initial seeding by diffusion, and the eventual development into mature forest by homeostasis, the latter is the more significant.

To illustrate examples of kinematic waves that depend only upon gradients and internal clocks and not upon diffusion, consider the effects of latitude. For instance the spring blossoming of trees is a wave moving north (in the northern hemisphere) and in autumn the onset of migration by birds is wave moving south.

The reader will easily recognise the existence of many visible primary waves of this nature in ecology, evolution, anthropology and sociology. However in this paper we are more concerned with *hidden* primary waves in developmental biology, that cause secondary waves of physical manifestations in cells, that in turn may cause morphogenesis. We shall therefore state and prove the theorem in this context.

5.   PRIMARY WAVES IN EMBRYOLOGY

We begin by enlarging a little on the meaning of the four hypotheses in developmental biology. There is no need to give precise definitions at this stage, because the terms are given precision by the way we choose to translate them into mathematics in the proof of the theorem in §8 below. Suppose that E is a multicellular mass of tissue. We are concerned with development of E during a particular time interval T.

I: Homeostasis means that each cell is in stable biochemical equilibrium, an equilibrium that may change with time.

II: Continuity means that at the beginning of T we can represent the chemical, physical and dynamical conditions in different cells by smooth functions on E (the conditions inside a particular cell are represented by the values of the functions at the centre of mass of that cell). In an embryo, where the tissue has developed from an egg by cleavage, the continuity is inherited from the original continuity in the egg, which was due to diffusion in the egg cytoplasm. Any slight discontinuities that arise later tend to be evened out by subsequent diffusion across the cell walls. In aggregates of cells like slime mold, continuity means that the cells have sorted themselves out according to continuous gradients during the aggregation process.

Continuity implies that neighbouring cells will follow nearby paths of development whenever possible. We shall prove that where a frontier stabilises this is not possible, and so across the frontier neighbouring cells will follow divergent paths of development, and large discontinuities will therefore arise.

III:Differentiation means that, whereas at the beginning of T there is only one type of cell (or, more precisely, a continuous variation amongst the cells), at the end of T there are two distinct types, and no continuous variation from one type to the other.

For simplicity we may assume that the tissue E is polarised, that is to say all variation takes place in one direction only (like the north-south line in the previous example in §4). Therefore for mathematical analysis it suffices to consider a 1-dimensional space interval S in that direction. Continuity means that at the beginning of T the cells vary continuously along S. Differentiation means that during T the cells at opposite ends of S develop continuously into different types. At the end of T, since there is no continuous variation between the two types there must be a frontier point in S separating the two types. This implies a frontier surface in E, separating the two types of tissue. If we can show that the frontier point in S moves, then this will imply that the frontier surface in E moves.

IV:Repeatability means that the development is stable, that is to say a qualitatively similar development will take place under sufficiently small perturbations of the initial conditions.

MAIN THEOREM: *Homeostasis, continuity, differentiation and repeatability imply the existence of a primary wave. In other words a frontier forms, moves and deepens, then slows up and stabilises, and finally deepens further.*

Therefore whenever a frontier forms, it first forms off to one side and then moves as a primary wave through the tissue before stabilising in its final position. Here by "final position" we mean the position relative to the underlying tissue, which itself may be undergoing morphogenetical movements. The theorem is illustrated in Figure 1, and the proof is given in §8 below.

Remark 1.   The theorem is qualitative rather than quantitative;

in other words it is a result invariant under diffeomorphisms of space and time. Therefore the theorem cannot predict the extent of travel of the wave, and so in applications the extent must always be taken as an extra hypothesis, and verified experimentally as in §7 below. It appears that some primary waves may travel a large distance, particularly those associated with morphogenesis. For example in gastrulation of some newts the ectoderm/mesoderm frontier travels from the grey crescent at latitude 40°S (see Figure 12) to its stabilisation position at 40°N, which is more than half the diameter of the blastula.

On the other hand the theorem does give quantitative predictions about the initial deepening, and the final stabilisation, of the frontier, because both these obey parabolic laws, and parabolicity is a diffeomorphism-invariant (see §9). These laws should furnish easily testable predictions.

Remark 2. The theorem gives no indication of whether the primary wave is visible or hidden. In embryology primary waves are generally hidden, in the sense of being experimentally undetectable at the time, because they probably consist of the switching on of gene systems, although their passage can sometimes be tracked in retrospect by the grafting experiment described in §7 below.

6. SECONDARY WAVES IN EMBRYOLOGY

The theorem gives no indication of whether or not a hidden primary wave will result in a visible secondary wave after a time delay; and if it does, the theorem gives no indication of the size of delay, nor of the type of secondary wave. These must depend upon extra detailed biochemical hypotheses about the particular systems that are switched on, what the long term effects these systems have on the cells, how these effects physically manifest themselves, and whether there is a resulting energy release. Broadly speaking there are three possibilities.

In order to describe the three cases it is necessary to be a little more precise about what we mean by the word *differentiation*.

Differentiation can be used in two senses, firstly the hidden determination of the cells into two types that takes place during the passage of the hidden primary wave, and secondly the subsequent development of physical difference between the two types that can be observed. In the hypothesis of the theorem in the last section we used differentiation in the first sense, because this was the best word to capture hypothesis III, the determination of distinct types. In this and the next section we use differentiation in the second sense, because this is the more normal usage. We now describe the three cases.

(i) *There is no secondary wave of energy release, and the speed of differentiation is slow compared with the speed of the wave.* No release of energy implies no morphogenesis. The slowness of differentiation implies that any secondary wave will probably be unnoticeable. Therefore the only visible effect will be the slow appearance of the frontier between the two types of tissue in its final position s_2. The original primary wave may never be noticed unless looked for.

(ii) *There is no secondary wave of energy release, and the speed of the differentiation is fast compared with the speed of the wave.* Again no release of energy implies no morphogenesis. However in this case the swiftness of differentiation implies that the frontier between the two tissues will appear before it has stopped moving, and so will present a visible secondary wave. The frontier may not necessarily first appear at s_1, because it may not yet be deep enough to notice, but it will appear at some point s_3, where $s_1 < s_3 < s_2$. It will then move towards s_2, deepening and slowing down according to a parabolic law. This means that near s_2 the speed of the wave s is proportional to $\sqrt{s_2 - s}$ (see §9, Corollary 3). The frontier then stabilises at s_2 and deepens further.

This is a commonly observed phenomenon, and is often described as *recruitment*. For example,* a developing insect eye starts with a few cells, and then enlarges by recruiting neighbouring epidermis cells. The word "recruitment" implicitly suggests that the experimenter should look for an "evocator" or an "organiser" emanating from the existing eye cells causing the recruitment.

---

*I am indebted to Peter Shelton for this example.

However if this expanding frontier of the eye were a secondary wave, then perhaps one ought to look for a hidden primary wave passing some hours before, without necessarily any organiser. In mammalian eyes the wave goes the other way: instead of expanding outwards the optic region shrinks in size.

(iii) *There is secondary wave of energy release.* In this case there will be a dramatically observable secondary wave causing morphogenesis. Geometrically the situation will be complicated by the fact that not only is the wave moving through the tissue, but the tissue itself is also moving and changing shape, with the former causing the latter. It is also likely that the energy release precedes differentiation. Therefore differentiation may (or may not, as in cases (i) and (ii) above) appear as another secondary wave after the morphogenesis. Therefore two misinterpretations are possible: Firstly, it may be geometrically unobvious to relate the visible secondary differentiation wave to the hidden primary wave, because the morphogenesis will have moved the issue around in between the two. Secondly, it may be tempting to conclude that the differentiation has been evoked by the morphogenesis, or by the new position of the tissue, rather than by the original hidden primary wave. Thus the experimenter's path may be strewn with pitfalls, unless he manages to uncover the primary wave, which possibility we now discuss.

7. EXPERIMENTAL DETECTION OF A PRIMARY WAVE

Sometimes the hidden primary wave may mark a loss in potentiality. In this case, the passage of the wave can be detected by a standard grafting experiment, provided that the tissue is suitable for grafting. For example one can detect the hidden mesoderm wave in amphibian gastrulation by this method*.

Suppose, as before, that the wave starts at s_1 at time t_1 and stabilises at s_2 at time t_2. At this stage it is hidden and so there is no detectable difference between the cells, but eventually at some later time t_3 differentiation will cause a physically observable

---

*I am indebted to P. D. Nieuwkoop for explaining this experiment to me.

difference that we indicate by shading the regions marked A, B in
Figure 2. At time t_1, although there is no difference, we can say
that a-cells are presumptive A-cells and b-cells are presumptive
B-cells. The wave travels from a towards b. The passage of the
wave past a cell is indicated by that cell switching from being a
b-cell to being an a-cell. The switch occurs essentially because of
homeostasis, as we explain in the proof of the theorem in the next
section. Therefore b-cells have the potentiality to develop into
either A-cells or B-cells. However a-cells may only have the
potentiality to develop into A-cells, and we suppose that this is the
case. Therefore the primary wave marks the loss in B-potentiality.

If we want to verify that the hidden primary wave has reached
position s at time t, then at time t graft two small pieces α, β from
just behind and just ahead of s onto b-tissue, well clear of the
wave, as shown in Figure 2. The β-cells are influenced by their

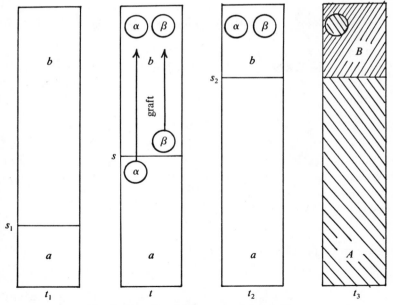

FIG. 2. Detecting a primary wave by grafting.

new position to remain b-cells, and so by time t_3 develop into B-cells, causing the β-graft to disappear. Meanwhile the α-cells have already switched and lost their B-potentiality and so by time t_3 the α-graft stand out as a patch of A against B.

8. PROOF OF THE THEOREM

We follow the conceptual ideas of Thom [18, 19]. As explained in §5 above it suffices to use a 1-dimensional interval of space S, transverse to the forming frontier. Let T be a time interval encompassing the development. Let $C = S \times T$ be the rectangle of space-time. Let X denote a manifold representing the states of a cell. One can envisage X as a bounded open subset of n-dimensional euclidean space R^n, where n may be very large (possibly several thousand). The coordinates $\{x_i; i = 1, 2, \ldots, n\}$ of a point $x \in X$ may represent not only the concentrations of the different proteins in the cell, and the rates of change of those concentrations, but also may include variables representing various physical characteristics of the cell, its membrane, the cell dynamics, etc.

Consider a cell at point $c \in C$. By Hypothesis I this cell is in homeostasis. We choose to translate homeostasis into mathematics by assuming that the biochemistry of the cell can be modelled by a gradient dynamical system on X

$$\dot{x} = - \operatorname{grad} V_c,$$

where $V_c : X \to R$ is a smooth function and R denotes the real numbers (see Remark 3 below for the meaning of this function). We choose to translate Hypothesis II, continuity, into mathematics by assuming that V_c can be chosen to depend smoothly on c. Therefore we have a function

$$V : C \times X \to R$$

given by $V(c, x) = V_c(x)$. We choose to translate Hypothesis IV, repeatability, into mathematics by assuming that V is generic*. Let

---

Generic means in general position, that is to say the map $c \to V_c$ maps C transverse to the natural stratificaton of $C^\infty(X)$. Generic V's are often dense in the space of all V's, and therefore both stable, and permissible to use as models.

$M \subset C \times X$ denote the set of stationary values of V, given by $\nabla V = 0$, where ∇ denotes the gradient with respect to X. Let G denote the closure of the subset of minima, which are given by $\nabla^2 V$ positive definite, where ∇^2 denotes the Hessian with respect to X. Then by smooth genericity, M is a smooth 2-dimensional surface in the $(n + 2)$-dimensional space $C \times X$, and G is a subsurface of M with boundary ∂G. Let $\chi : M \to C$ denote the map induced by the projection $C \times X \to C$. By Thom's classification theorem of elementary catastrophes [18], the only singularities of χ are fold curves and cusp points, since V is smooth and generic. The boundary ∂G consists of fold curves and cusp points.

By homeostasis the state of cell at c is at a minimum of V_c, and therefore represented by a point

$$\sigma(c) \in G \cap \chi^{-1}c.$$

Therefore σ is a section of χ,

$$\chi \downarrow \overset{G}{\underset{C}{\Big)}} \sigma.$$

In other words $\chi\sigma = 1$. The interesting point is that, whereas χ is smooth (induced by projection), σ may be forced by χ to be discontinuous. We now use Hypothesis III, differentiation, to analyse this discontinuity and show, using I and IV again, that it implies the primary wave.

Let $S = [s_0, s_3]$, $T = [t_0, t_3]$. Let us analyse the continuity of σ on the boundary ∂C of $C = S \times T$. Firstly σ is continuous along the side $s_0 \times T$ because a-cells are developing smoothly into A-cells. Similarly σ is continuous along $s_3 \times T$ because b-cells are developing smoothly into B-cells. Next σ is continuous along the side $S \times t_0$, because, by Hypothesis II, continuity, we may assume that at the beginning the tissue is continuous.

Finally σ cannot be continuous along the side $S \times t_3$, because, by Hypothesis III, differentiation, at the finish two distinct types of cells A, B have developed, with no continuous variation between them. By Hypothesis II, continuity, A must spread continuously from one end, and B from the other, towards some point of discontinuity, which is, as yet, undetermined.

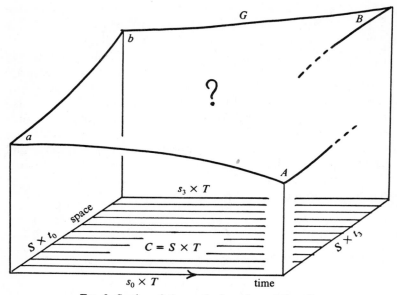

FIG. 3. Section of G over the boundary of $S \times T$.

Figure 3 shows the partial section of G over ∂C that must exist ready to receive the map $\sigma : \partial C \to G$. Now comes the problem of extending σ to the interior of C. First we ask the simpler question: what singularity must the map $\chi : M \to C$ have over the interior of C? Since $M \supset G$, M must extend the section over ∂C shown in Figure 3. Therefore by the classification theorem M must have at least one cusp singularity over the interior of C. A single cusp would be sufficient. Moreover a single cusp is qualitatively the simplest solution of the extension problem, and we can justify the simplest solution by again appealing to Hypothesis II, continuity; in other words we assume minimal discontinuity subject to Hypothesis III, differentiation.

In Figure 4 we illustrate two examples of a surface M over C, each with one cusp, and each extending the given section over ∂C. In each case the shaded subsurface indicates M-G (representing saddle-points of V, in other words unstable equilibria of the biochemical dynamic, and so not realisable by homeostasis). If we

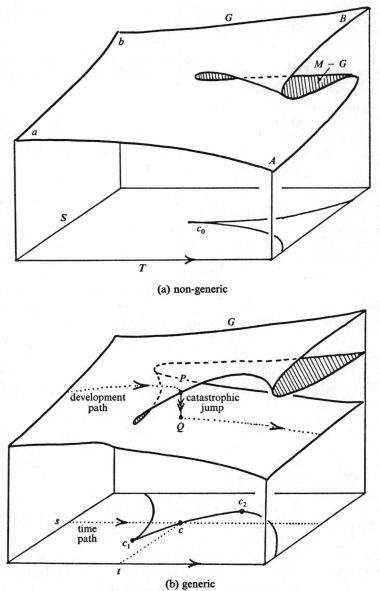

(a) non-generic

(b) generic

FIG. 4. The graph G of homeostatic states.

ignore the product structure of space-time $C = S \times T$, then the two pictures are qualitatively equivalent. However if we take note of the product structure, then Figure 4a is exceptional because the time-axis at the cusp point c_0 coincides with the cusp-axis. Hypothesis IV, repeatability, forbids this exceptional situation and so Figure 4a is ruled out.

The reader may ask why do we choose the rather elaborate Figure 4b, and so we must explain. Firstly the direction of the cusp-axis must have a non-zero S-component (to avoid the fault of Figure 4a) and a positive T-component, because the frontier between the differentiated tissue must occur inside the cusp. Probably in most cases the T-component will be greater, but to emphasise both components in Figure 4b we have drawn them approximately equal, in other words in the perspective drawing of C we have drawn the cusp-axis inclined at $45°$ to the time-axis at c_1. Secondly, as the two branches of the cusp widen out, there will be a unique first point c_2 (on the branch for which t is greater) where the tangent is parallel to the time-axis. These two points c_1, c_2 will mark the beginning and ending of the primary wave, as we shall now prove. We have merely drawn Figure 4b so as to emphasise the qualitative features of these two points.

We now plot the development of each cell by lifting its time-path in C up onto the graph G. Since G is the graph of homeo-static states, the lifted path will represent how the state of the cell changes, in other words will represent its *development-path* (see Figure 4b).

Since homeostasis is represented by a differential equation $\dot{x} = -\nabla V$, the development path will be a continuous path on G, held continuously in stable equilibrium by the differential equation, unless the path happens to cross ∂G. Therefore, in the language of Thom [18], the changing state will obey the Delay Rule. Now ∂G is the fold curve of M lying above the two branches of the cusp. Suppose that the time-path of the cell at s crosses the branch $c_1 c_2$ of the cusp at the point c at time t (see Figure 4b). Then the development-path of s will cross ∂G at the point P above c, and at this point the homeostatic stability breaks down, because the corresponding minimum of V_c has coalesced with a saddle (represented by $M - G$), and disappeared. Consequently the homeostatic differential equation comes into play and carries P rapidly to Q, which is the unique new stable equilibrium on G above c, in whose basin of attraction P lies (see [26]). The rapid change of state from P to Q caused by homeostasis is called a *catastrophe*, or *catastrophic jump*; this is the moment when the b-cell at s switches into an a-cell. What we have previously vaguely referred to as "switching on of gene systems" is represented mathematically by the fast flow from P to Q along an orbit of the differential equation in X representing homeostasis. That was why in the introduction and in §7 above we remarked that the switch occurs essentially because of homeostasis. Therefore at time t the cell s marks the frontier between a–cells and b–cells, namely the position of the wave. The depth of the frontier is represented by the length PQ. The catastrophe occurs in all cells lying in the interval $s_1 < s < s_2$ during the period $t_1 < t < t_2$, as their time-paths cross the cusp branch $c_1 c_2$, and this defines the primary wave. Therefore $c_1 c_2$ determines the track of the primary wave in space-time. Meanwhile cells for which $s < s_1$ suffer no catastrophe, but develop from a-cells into A-cells along a smooth development-path. Similarly cells for which $s > s_2$ develop smoothly from b-cells into B-cells. The various development paths are shown in Figure 5.

From Figure 5 we can read off the qualitative features required in the statement of the theorem, as follows:

At time t_1 the frontier first forms at s_1.

Between t_1 and t_2 the frontier moves from s_1 to s_2 and deepens.

As the time approaches t_2 the frontier slows up, and approaches s_2.

At t_2 the frontier reaches s_2, and stabilises.

After t_2 the frontier deepens further.

This completes the proof of the theorem. Before we proceed to quantitative features of the theorem we make four remarks.

Remark 1. Sometimes the wave does not begin in the middle of the tissue, but on the boundary of the tissue, so that the tissue appears to "grow into" the frontier. This seems to be the case with slime mold (see §15 below), and with the development of chicken wings, for instance. In this case the mathematics is simpler because the space-time track of the tissue does not cross the cusp

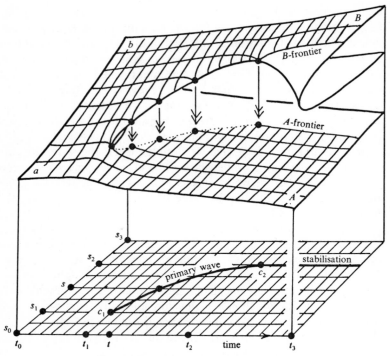

FIG. 5. Development paths of cells on G.

point, but only the fold curve. In fact there may not necessarily be any cusp point at all. In Figure 5 the slime-mold grex would be represented by $[s, s_3]$ with tip at s and tail at s_3. At time t the wave would begin at the tip, and then proceed along the grex to stabilise at s_2 at time t_2. The front part $[s, s_2]$ eventually develops into stalk-cells A, and the back part $[s_2, s_3]$ into spore-cells B.

Remark 2. Not all frontiers are formed by primary waves of this type, because in some cases our translation of the four hypotheses into mathematics may not be valid. For example in the gastrulation of birds and mammals, or in mixing experiments [1, 10], the frontier is caused by migration of different types of cells, sorting themselves out, whereas we have assumed that the cells stay more or less in the same place relative to one another in the tissue. However in some of these cases a primary wave may already have taken place in some underlying gradient, and the migration of cells up or down the gradient may be merely a secondary wave.

Remark 3.   There was one drastic simplification that we made in the proof of the theorem, in the way that we chose to translate homeostasis into mathematics. It may well be reasonable to represent homeostasis by a dynamic D on X, but it is not obvious that D should be a *gradient* dynamic, $\dot{x} = -\nabla V$. In special cases V may represent some potential energy that is minimised, and then it is reasonable. But in general D may be non-gradient, particularly when the cell contains biological clocks. Even then in some cases it is technically possible to reduce D to the gradient case, by choosing V to be a Lyapunov function for D (see [18, 26]). However in other cases this may not be possible; frontiers arising from, or associated with turbulence, for instance, would probably not behave so simply.

Remark 4.   We have drawn Figures 3, 4, 5 as if X were 1-dimensional, and as if

$$M \subset C \times R \subset R^3.$$

In fact this is not true because X is an open subset of R^n, where n may be very large, and therefore more precisely

$$M \subset C \times X \subset R^{2+n}.$$

However this does not alter the fact that M is 2-dimensional surface, and therefore our diagrams are indeed rigorous pictures of the map $\chi: M \to C$. Moreover Thom's classification theorem [18] can be modified [26] in an important manner for this context, as follows:

If $\chi: M \to C$ has a cusp catastrophe, then in the neighborhood of that point we can choose a map $\pi: X \to R$ such that

$$1 \times \pi: C \times X \to C \times R$$

throws M diffeomorphically onto the surface pictured in Figure 4b. Moreover π can be chosen to be the projection of X onto one of the given axes of R^n, that measure concentrations etc. In fact we can choose any axis not perpendicular to the tangent to ∂G at σc_1. Let us call the chemical or physical property that this axis measures, a *morphogen*. Then the vertical axis in Figures 3, 4, 5 measures the morphogen. The morphogen need not be unique, and may only be an artifact. But if the morphogen is easy to measure, then it may be useful for experimental predictions. It is remarkable that Thom's theorem guarantees the existence of a morphogen for each developing frontier.

9. QUANTITATIVE ASPECTS OF THE THEOREM

From Figure 5 we can deduce some quantitative estimates about primary waves. The estimates are computed to first order in small quantities, and are therefore only accurate near the beginning and the end of the wave.

COROLLARY 1: *Initially, when the frontier first forms, it is moving at constant speed.*

Proof: The path of the wave is the branch c_1c_2 of the cusp, which, near c_1, to first order, can be replaced by the tangent at c_1.

COROLLARY 2: *Initially, when the frontier first forms, its depth increases by a square-root law, in other words the depth of the frontier is proportional to $\sqrt{t - t_1}$ (and hence also to $\sqrt{s - s_1}$).*

Proof: At time t the depth of the frontier is equal to the catastrophic jump PQ in Figure 4. Therefore we must compute PQ. Choose origin 0 at σc_1, the point of M over the cusp point c_1. Choose two axes ξ, η at 0 as follows: ξ is measured along the tangent at 0 to ∂G in X, oriented towards P, and η is measured along the tangent at c_1 to the cusp in C. Let K denote the (ξ, η)-plane. Then K is the osculating plane of ∂G at 0. Therefore, by genericity, and ignoring third order terms, ∂G lies in K and has equation $\eta = k\xi^2$, where $k > 0$. Therefore P satisfies $\xi = + \sqrt{\eta/k}$. Meanwhile Q satisfies $\xi = -2\sqrt{\eta/k}$, because M is the diffeomorphic image of a cubic surface, namely the canonical cusp catastrophe. Therefore $PQ = 3\sqrt{\eta/k}$. But η is proportional to $t - t_1$, and hence PQ is proportional to $\sqrt{t - t_1}$, as required.

Remark.   The initial position and movement of a developmental wave may be difficult to observe, because of the initial shallowness of the frontier. However it might be possible to find the initial position by using Corollary 2 to extrapolate backwards (and hence find the organising centre, if the wave happens to emanate from a point).

Note that Corollary 1 also remains true for any secondary wave. However we should not necessarily expect Corollary 2 to apply to a secondary wave, because the two waves are of a totally different nature: the primary wave marks the frontier between two diverging types of tissue, whereas the secondary wave makes the onset of a secondary effect within one type of tissue. Therefore the initial movement of the primary wave may be difficult to observe, whereas that of the secondary wave may be easy to observe—for instance the first invagination in gastrulation (see §10 below).

COROLLARY 3:   *Eventually, just before the frontier stabilises, it slows down parabolically. In other words $(s_2 - s)$ is proportional to $(t_2 - t)^2$, and the speed is proportional to $(t_2 - t)$.*

Proof: By genericity the curve $c_1 c_2$ touches the time axis at c_2 with quadratic tangency. Therefore near c_2, ignoring third order terms, the curve has the equation

$$(s_2 - s) = h(t_2 - t)^2,$$

where $h > 0$. The speed is given by differentiating:

$$\dot{s} = 2h(t_2 - t).$$

This completes the proof of Corollary 3.

Remark. Corollary 3 should be easy to observe and verify. If there is no morphogenesis, then the same result will hold for any secondary wave. Therefore in a recruitment phenomenon, for instance, the parabolic law might provide a good test to distinguish whether it was a secondary wave or the result of entrainment.

If there is morphogenesis, then the displacement of cells relative to one another may upset the parabolic law for the secondary wave, but Corollary 3 may nevertheless yield other quantitative predictions–see for example the estimate for stalk diameter in the slime mold fruiting body, in §16 below.

Energy release. Some morphogenetical movements begin slowly, and build up to a recognisable climax before finally dying down. This can be seen most clearly in time lapse films, and the reader is especially recommended to see the two Göttingen films of Luther [11] on gastrulation and neurulation, and Gerisch [7] on slime mold. For instance, in [11] gastrulation begins by invaginating slowly, then the tissue begins to roll over the dorsal lip, then pours over the entire circle of blastopore lip, until it eventually slows down, and the blastopore gently closes. Similarly neurulation begins with the neural folds appearing slowly, then they rear up and the neural tube snaps shut in the middle, and the closing process runs towards both ends, which eventually seal themselves more gently. In [7] the slime mold fruiting body heaves itself slowly off the ground, then accelerates and rises rapidly up its stalk, then slows down, and eventually the knob at the top gently disappears.

We shall now show, by computing the energy released, that this is the normal pattern for a morphogenetical movement arising from a secondary wave.

Note that in the films it can also be observed that the gastrula gives a final heave before the blastopore closes, and the slime mold fruiting body gives a hiccup halfway up; but these are subsidiary frictional effects that we shall explain later.

Assuming that the primary wave begins at time t_1 and ends at time t_2, let

$V(t) = $ speed of primary wave at time t, $t_1 < t < t_2$;

$A(t) =$ area of wave-front at time t, $t_1 < t < t_2$;

$e(\tau) =$ rate of energy-release by a cell at time-interval τ after the primary wave has passed it. We may suppose that $e(\tau) = 0$ outside an interval $\delta_1 < \tau < \delta_2$, where $0 < \delta_1 < \delta_2$, and where δ_1 is the delay between the primary and secondary waves, and $[\delta_1, \delta_2]$ the period during which energy is released by the secondary effect.

LEMMA 1: *The total rate of energy release at time t, where $t_1 + \delta_1 < t < t_2 + \delta_2$, is given by*

$$E(t) = \int_{t - \delta_2}^{t - \delta_1} A(\tau)V(\tau)e(t - \tau)d\tau.$$

Proof: The number of cells crossed by the primary wave in the interval $[\tau, \tau + d\tau]$ is $A(\tau)V(\tau)d\tau$, and by the time t each of these is still releasing energy at the rate $e(t - \tau)$. Integrating gives the lemma.

In Figure 6 we sketch the qualitative shape of the graph of E. The assumptions on which the sketch is based are as follows: V is initially constant near t_1, by Corollary 1, and eventually decreases linearly to zero at t_2, by Corollary 3. If we assume that the primary wave emanates from a point, and that the wave-front initially expands linearly, then we deduce that A starts from zero at t_1 and initially increases parabolically, in other words proportional to a square law. Eventually A becomes constant as the frontier stabilises at t_2. We assume the secondary effect e starts suddenly at δ_1, and then decreases linearly to zero by δ_2. It can be shown, by using the lemma to integrate these assumptions, that E both begins and finishes proportional to cube laws.

Summarising: this characteristic pattern of energy release during a morphogenetic movement might provide a useful clue that the movement was the secondary effect of an earlier hidden primary wave. George Oster suggests that the energy released by individual cells might be observed by measuring the heat loss microcalorimetrically.

234   E. C. Zeeman

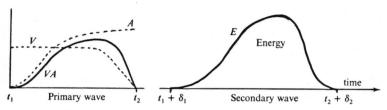

FIG. 6. Graph of energy released by a secondary wave.

The ripple ahead of a wave.   Consider a fixed time t during the primary wave, $t_1 < t < t_2$. The state of the tissue is obtained by lifting the section $S \times t$ of C up onto G in Figure 5. Consider the variation in the state of the cells as the frontier is approached from either side. On the a-side the state is approximately constant, and so the cells are homogeneous, but on the b-side the variation is parabolic as the frontier is approached.

In some cases this phenomenon might be repeated visibly in a secondary wave. For instance, if we had a situation as in §6(ii) above, where the secondary wave was differentiation, then the phenomenon might be visible as a slight ripple ahead of the wave. For example, suppose the expanding frontier of the insect eye were a secondary wave. Then inside the frontier the already recruited eye cells should appear relatively homogeneous, but outside the frontier, the epidermis cells just about to be recruited might show physiological signs of the impending recruitment. In mammalian eyes the effect might appear on the inside of the frontier, because the wave goes the other way.

REFERENCES

1. B. I. Balinsky, *An Introduction to Embryology*, 2nd ed., Saunders, Philadelphia and London, 1965.

7. G. Gerisch, *Dictyostelium discoideum* (*Acrasina*) *Aggregation und Bildung des Sporophors* (Institut für den Wissenshaftlichen Film, Göttingen), Film E 631 (1963).

8. T. Gustafson and L. Wolpert, "The forces that shape the embryo," *Discovery*, **22** (1961), 470–477.

10. C. K. Leach, J. M. Ashworth, and D. R. Garrod, "Cell sorting out during the differentiation of mixtures of metabolically distinct populations of Dictystelium discoideum," *J. Embryol. Exp. Morph.*, 29, **3** (1973), 647–661.

11. W. Luther, *Entwicklung des Molcheies* (Institut für den Wissenshaftlichen Film, Göttingen) Film C. 939 (1967).

18. R. Thom, *Stabilité Structurelle et Morphogénèse*, Benjamin, New York, 1972.

19. ———, A global dynamical scheme for vertebrate embryology, (A. A. A. S. 1971, Some Mathematical Questions in Biology, VI), *Lectures on Maths. in the Life Sciences*, vol. 5, Amer. Math. Soc., Providence, 1973, 3–45.

22. C. H. Waddington, *The Strategy of the Genes*, Allen & Unwin, New York, 1957.

24. E. C. Zeeman, Differential equations for the heartbeat and nerve impulse, *Towards a Theoretical Biology* 4, ed., C. H. Waddington, Edinburgh University Press, 1972, 8–67.

26. ———, Gradients and catastrophes in developmental biology, (manuscript).

THE LARGE SCALE STRUCTURE AND DYNAMICS OF GENE CONTROL CIRCUITS: AN ENSEMBLE APPROACH

Stuart Kauffman

1. INTRODUCTION

The continuing success of molecular biology in discovering the mechanisms controlling the expression of individual or small groups of genes is bringing to the foreground questions about the large scale organization and dynamic behavior of cellular control systems, and how best to discover their nature. Current estimates of the number of genes, both control and structural, in a higher metazoan, range from 40,000 up to about 1,000,000. The uncertainty reflects in part the difficulty in assessing the role and extent of redundant DNA. Although it may turn out that these genes are organized into very simple control circuits, as suggested by Ohno (1971), that possibility is no certainty. Biologists should consider whether there may be some systematic way to use what is known to generate hypotheses about the likely organization of large systems of genes, how much of the cellular control system we can reasonably expect to know, and what might be explained with what we expect to know.

It is reasonable to expect continuing discovery of the major kinds of molecules playing control roles, their mechanisms of action, and of small scale, local properties of the organization of these molecular processes into control systems. In particular, it is proving possible to discover which molecular processes directly control a given process—the immediate control connections in the system; and how variations in the controlling variables affect the controlled process. Where these processes are grouped into reasonably simple intermediate sized control circuits, the properties of those circuits should be discoverable. This is already happening with bacteriophage lambda. However, it seems unreasonable to expect to discover virtually all the control relations among 40,000 or 1,000,000 genes. Therefore, we should consider ways to construct an adequate picture of the architecture of cell control systems whose full details may never be directly known. In addition, incomplete knowledge of those control systems poses the critical problem that there are likely to be dynamic properties of central biological importance which depend in some way on large portions or on the whole organization of the control system, not on small isolatable fragments of it. A number of well-known cellular dynamic properties probably reflect the overall organization of cellular control systems. Among these are: (1) the pattern of gene activities corresponding to any one cell type, in any organism, must be restricted to a relatively small number of combinations of gene activity through which the cell "modulates" in its ongoing activities; (2) any organism possesses a particular number of stable distinct cell types; (3) during ontogeny, any cell type differentiates directly into rather few other cell types, although it may eventually differentiate into many by repeated branching differentiations. Large scale properties such as these presumably reflect and imply something about the overall architecture of cellular control systems whose design we wish to learn. If there are large scale dynamic properties of interest depending upon control systems whose full details are, and may remain, unknown, it is appropriate to assess how we can begin to link them to the kinds of small scale properties we expect to know.

One approach is to characterize any known small scale properties of the organization of cellular control systems, such as specifying the typical number of variables controlling any process and specifying the ways variations in the controlling processes affect the controlled processes. Specification of such small scale, local properties should be useful in two ways: (1) the local properties form the basis for hypotheses about the organization of larger control circuits; (2) the implications of the small scale properties for the large scale dynamic behavior of cellular control systems can be assessed. Systematic use of such local characteristics for both these purposes can be made by constructing a set of all the possible large control systems, each member of which is built using only those small scale properties. This set, or ensemble, represents the class of hypotheses about the total architecture of cellular control systems implied by known small scale properties of the organization. Examination of the typical, or average "wiring diagram" of ensemble members will allow hypotheses about the most probable kinds of intermediate and large control circuits which may be expected from small scale properties we already know. Examination of the typical large scale dynamic behaviors of the ensemble members will allow us to assess the most probable large scale behaviors of cellular control systems having the known small scale properties. The primary purpose in characterizing small scale properties of cell control systems and constructing an ensemble of possible control systems is to examine the implications of known small scale features for probable large scale properties, rather than directly to help learn about molecular mechanisms or the small scale properties themselves.

In this article I present evidence for three claims: (1) nearly all known regulated genes and processes are controlled according to a very small, similar, and simple class of rules; (2) this small scale property appears sufficient to account for the known large scale dynamic behaviors of cellular control systems described above; (3) restriction to this class of control rules predicts the existence of simple but powerful intermediate size regulatory circuits with properties useful for directing differentiation; lambda phage appears to contain such a circuit.

2. THE *Lac* OPERATOR

Consider the *lac* operator of *Escherichia coli*. If bound by repressor (Jacob & Monod, 1963) it prevents transcription of the adjacent structural genes, Z, Y, A. Binding of the operator by repressor is controlled by the presence of repressor molecules, and lactose, the inducer. A transgalactosidation derivative of lactose, allolactose (Zubay & Chambers, 1971; Burstein, Cohn, Kepes & Monod, 1965; Muller-Hill, Rikenberg & Wallenfels, 1964) binds to a site on the repressor molecule, weakening the repressor operator bond and removing the repressor. Binding of the operator is dependent on the concentrations of repressor molecules and allolactose. In the absence of allolactose, binding of one operator saturates as repressor concentration increases. At a fixed maximal level of about 12 repressor molecules per cell (Bretscher, 1967), binding of lactose derivative to repressor saturates as lactose level increases (Riggs, Newby & Bourgeois, 1970).

To understand the behavior of the operator locus, it is convenient to examine its response to saturating, and minimum concentrations of its controlling molecular variables. In those cases the operator can be bound only if repressor is present in saturating levels, and no allolactose is present. The operator cannot be bound if there is either (1) no repressor or (2) saturating levels of allolactose. The striking feature of this process is that each control variable, acting alone, can determine one of the two possible states of the operator, regardless of the concentration of the other regulatory molecule. Absence of repressor, or high levels of allolactose, each alone, assures the operator cannot bind repressor. Both must be co-ordinated to assure the operator *is* bound. I aim to show that it is typical of regulated genes and processes that at least one control variable has a state which determines the outcome of the regulated process regardless of the states of other regulatory variables. This property defines a class of control rules I term canalizing functions.†

3. CANALIZING FUNCTIONS

The simplest way to describe the behavior of a regulated locus like an operator at saturating and minimal levels of its control variables, is to consider it an on-off device and use logical

---

†In previous publications I called "canalizing" functions "forcible" functions.

(Boolean) algebra. Let "$Op = 1$" mean repressor is bound to the operator, "$Op = 0$" mean the operator is unbound, "repressor = 1" mean repressor is present in saturating levels and "repressor = 0" mean repressor is absent. The behavior of the *lac* operator can be described by a table listing all possible combinations of the states of the molecular variables controlling it, and its subsequent response. Any such table is a Boolean function.

Boolean functions may be classified by the minimum number of controlling variables whose states must be specified in order to determine one specific state of the regulated processes. For example, one variable suffices to determine one state of the function, shown in Table 1, governing the *lac* operator; repressor = 0 determines $Op = 0$, regardless of the state of allolactose, the other control variable. Also, specifying allolactose = 1 by itself determines $Op = 0$ regardless of the state of repressor. Determining the state $Op = 1$ requires co-ordination of both variables; allolactose must equal 0 and repressor must equal 1. I shall call a function canalizing if at least *one* control variable has *one* state which, by itself, can determine *one* state of the regulated processes. In this sense, the *lac* operator is canalizing, since allolactose = 1 determines $Op = 0$ regardless of the state of repressor. Only one state of a canalizing variable, the *canalizing state*, guarantees that the recipient goes to a specific state; the other state of that variable guarantees nothing. Allolactose = 1 guarantees the operator is unbound; allolactose = 0 guarantees nothing, for the state of the operator still depends on the state of the repressor. The *canalized state* of a process governed by a canalizing function is that state which can be determined by a single regulatory variable; for the *lac* operator the canalized state is $Op = 0$. A process controlled by K variables may have up to K canalizing variables. The *lac* operator has two canalizing variables since either allolactose = 1 alone, or repressor = 0 alone guarantees $Op = 0$. By contrast, consider a hypothetical structural gene (ST) with three distinct adjacent operators, where repression of transcription requires binding of any two or more operators. Since determining either the state 0 or the state 1 for transcription would require specifying the states of at least two control variables, this function is not canalizing. Consider two hypothetical promoters regulating one structural gene (*St*), in which transcription occurs if either, but not both, promoters are bound by polymerase (Table 3). This Exclusive Or function is not canalized by either control variable, for no state of either promoter alone can determine whether transcription occurs or fails.

TABLE 1
The lac operator, a canalizing function

Allolactose	Repressor	Op
0	0	0
0	1	1
1	0	0
1	1	0

TABLE 2
A non-canalizing function

Op_1	Op_2	Op_3	ST
0	0	0	1
0	0	1	1
0	1	0	1
0	1	1	0
1	0	0	1
1	0	1	0
1	1	0	0
1	1	1	0

TABLE 3
A non-canalizing function

P_1	P_2	St
0	0	0
0	1	1
1	0	1
1	1	0

Boolean functions are a convenient idealization. Gene activities and other biosynthesis processes are more realistically described by Michaelis–Menten or co-operative sigmoidal binding curves. These binding curves are typically monotonically increasing or decreasing functions of one or more controlling variables, and bounded due to saturation. The concept of canalization generalizes naturally to these functions (Kauffman, 1970; Newman & Rice, 1971). For example, Fig. 1 depicts an arbitrary catalytic component subject to facilitation and inhibition of activity. Any level of inducer (I) suffices to determine that the activity of the catalyst ($\overset{\circ}{X}$) will be constrained to *one* side (below) of a boundary. Decrease of I determines that the constraint boundary will decrease. Increase of I alone determines nothing, for it can always be compensated for by increase of repressor. Decrease of I, or increase of R, canalizes catalytic activity toward 0. For simplicity, the Boolean notation will be used below.

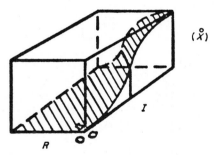

FIG. 1. A continuous monotonic bounded function for activity ($\overset{\circ}{X}$) as a function inducer (I) and repressor (R) concentrations.

4. BACTERIAL AND PHAGE GENES ARE GOVERNED BY CANALIZING FUNCTIONS

I shall consider the direct control variables for a structural gene in phage or bacteria to be "*cis*" acting sites within the same chromosome such as promoter, operator and termination sites. These *cis* regulatory sites are usually directly controlled by diffusible "*trans*" acting substances such as repressors, inducers, Rho, etc.

The proximate regulating loci for the *lac* structural genes in *E. coli* are the promoter and the operator (Zubay & Chambers, 1971). The operon structure is *POZYA*, in which P = promoter, O = operator, Z, Y, A are the structural genes and the five loci occur in the order *POZYA* on the chromosome. Let "$P = 1$" represent polymerase bound to promoter, and "$Op = 1$" represent operator blocked. Transcription occurs only if $P = 1$, $Op = 0$, Table 4.

TABLE 4

P	Op	ZYA
0	0	0
0	1	0
1	0	1
1	1	0[†]

This is the *Not If* Boolean function in which both control variables are canalizing, since $P = 0$ or $Op = 1$ each alone determines $(ZYA) = 0$. Note that $P = 1$ or $Op = 0$ alone is unable to determine a state of transcription. Only one state of a canalizing control canalizes the behavior of the target. The *lac* operator, as described above, can be canalized to $Op = 0$ by each of its two control variables.

Binding of polymerase at the *lac* promoter is regulated by polymerase core and sigma factor (Zubay & Chambers, 1971). In addition, like other catabolite repressible operons, cyclic AMP and the catabolite gene activation protein, CAP, are required for activation (De Crombrugghe *et al.*, 1971). Binding of repressor at the operator (Zubay & Chambers, 1971) may sterically hinder binding of polymerase to the promoter. All five control variables are canalizing: absence of each of the first four, or presence of

---

[†]The input state with promoter and operator simultaneously bound may be sterically impossible (Zubay & Chambers, 1971). If state 0 is assigned to transcription for that forbidden input state, the function is Not If, with two canalizing variables; if state 1 is assigned, transcription becomes a trivially canalizing function of the promoter alone.

bound operator alone determines polymerase fails to bind to promoter.

The probable sequence of genes of the arabinose operon on *E. coli*'s chromosome is *COIBAD* (Zubay, Gielow & Englesberg, 1971). *B*, *A* and *D* are structural genes, *O* is an operator, and *I* probably functions as a promoter: *C* product, regulator protein, probably can exist in two forms, P_1 and P_2, which attach respectively to *O* and *I*. The complex P_1 at *O* inhibits transcription of *BAD*. P_2 at *I* is required for transcription. L-arabinose is a specific effector, probably binding to the regulatory protein and stabilizing P_2 over P_1. Activation requires CAP and cyclic-AMP. CAP probably interacts with a locus in or near *I*.

Like *lac*, the arabinose structural genes are governed according to the canalizing Not If function. Each *cis* control variable alone can determine transcription fails (Table 5).

TABLE 5

O	I	BAD
0	0	0
0	1	1
1	0	0
1	1	0

The Ara *O* locus is controlled by L-arabinose and *C* protein. Let $Op = 1$ mean the *O* locus is bound. The function, Not If, is canalized by both molecular controls, since $C = 0$, or L-arabinose $= 1$ each suffices to determine $Op = 0$, Table 6.

TABLE 6

C	L-arabinose	Op
0	0	0
0	1	0
1	0	1
1	1	0

The presumptive molecules controlling the I locus are polymerase core, sigma factor, CAP, cyclic-AMP, C protein and L-arabinose. All six variables are canalizing, since absence of each determines that polymerase is not bound at I.

Bacteriophage lambda's right operator, O_r, is regulated by lambda repressor, and some metabolic signal induced by ultraviolet irradiation and other stimuli (Ptashne, 1967; Ptashne, 1971), which appears to render the repressor unable to bind O_r. The function is Not If, canalized by each variable to the size $O_r = 0$.

Lambda's left operator, O_1, is more complex. Lambda repressor, the product of gene C_1, binds to it and is removed, presumably by the same substance (X) which frees O_r during lytic induction. However, the product of lambda gene tof represses leftward transcription by binding at O_1 even in the presence of X (Eisen & Ptashne, 1971; Kumar, Calef & Szybalski, 1970; Szybalski $et\ al.$, 1970). O_1 is governed by the following function:

TABLE 7

tof	C_1	X	O_1
0	0	0	0
0	0	1	0
0	1	0	1
0	1	1	0
1	0	0	1
1	0	1	1
1	1	0	1
1	1	1	1

This function has a canalized state, $O_1 = 1$, and tof is a canalizing control variable, since $tof = 1$ determines $O_1 = 1$ regardless of the states of C_1 or X. Neither remaining control variable is canalizing because no state of C_1 or X alone can determine $O_1 = 1$.

Processes governed according to canalizing functions are not limited to direct gene activity. In $E.\ coli$, the arginine operon may be regulated by blocking translation (Vogel, McLellan, Hiroven &

Vogel, 1971). The exact nature of the repressing complex is not yet known, but it is clear that translation block requires the Arg R product as aporepressor, and arginine as co-repressor. Absence of either control molecule vetoes the block. Histidine (Brenner & Ames, 1971) may also involve translational control and appears governed by a canalizing function.

Since canalizing functions become a very small fraction of the possible control rules as the number of controlling variables increases above three (Kauffman, 1970), and Tables 2 and 3 consider possible regulated genes which are not controlled by canalizing functions, the assertion that regulated genes are commonly controlled according to canalizing functions is not trivially true.

The examples given were chosen only because data concerning them are clear. Numerous examples can be found in the literature on bacterial and phage gene regulation [*Metabolic Pathways,* vol. 5 (1971); Umbarger (1969); *The Bacteriophage Lambda* (1971)]. Difference in molecular processes between bacteria and higher cells weakens generalization based on detailed molecular mechanisms. However, functional properties of control organization may be less dependent on their molecular embodiment. While examples of canalizing functions were drawn from phage and bacteria, the property of canalization depends only on the properties of monotonicity and saturation in the binding curves of ligands controlling activity of catalytic elements. Langmuir and sigmoidal binding curves are common and canalization may occur at many distances from direct gene activity. It appears difficult to find many examples of controlled metabolic and genetic processes which are not regulated according to canalizing functions. For the theory I shall develop, it is sufficient if most (but not necessarily all) processes with more than one control variable utilize canalizing functions.

5. LARGE SCALE DYNAMICS

The purpose in trying to establish the class of phenomenological control rules governing gene activities is to characterize small scale properties of the organization of cell control systems. Such local properties at best constrain hypotheses about large scale organization to the enormous class of possible large control systems built consistent with the local properties. The implications of these small scale properties for the large scale dynamic behavior of control systems can be assessed by constructing an ensemble of all the possible large control systems in which each member is built using only the local properties, and asking whether the ensemble possesses "typical", or expected large scale dynamic properties which occur in the vast majority of the ensemble, but which do not occur in systems built without these local constraints.

The examination of regulated bacterial processes above seems to warrant two conclusions: most genes are probably regulated according to canalizing functions; most are directly controlled by rather few, one to six, other processes. The latter implies that the mean connectivity of the control system is low.

To assess the implications of these local properties, one would ideally use realistic Michaelis–Menten, sigmoidal, or other continuous kinetic equations to express the kinds of canalizing functions found, and explore the behavior of the ensemble of systems built using them. The ideal is not directly approachable, for no adequate techniques exist to examine large systems of such nonlinear differential equations. However, the question may be approached by substituting the idealization of a regulated process like gene activity as a binary, on-off device. Very large systems of such binary genes are easily studied by simulation.

To discover the typical dynamic behavior of this ensemble of systems requires examining the behavior of systems sampled at random from that ensemble. This may be done by building control systems in which each model gene is assigned at random one to five or six other genes as its control variable, and each gene is assigned at random one of the possible canalizing functions on its control variables. Once built, any such network is fixed in structure and is a random sample from the ensemble of possible control systems. The dynamic behavior of members may be compared to typical members of an ensemble of control systems built without using canalizing functions. Since I have reported results of such studies elsewhere (Kauffman, 1970, 1969a, b, 1971; Glass & Kauffman, 1972, 1973) I will only summarize.

In a network of 10,000 binary genes, each governed by a

non-canalizing function of many other model genes, the following typically occurs:

(1) The net has $2^N = 2^{10,000}$ ($\approx 10^{3000}$) distinct states comprised by each possible combination of gene activities for the N genes. When released from an initial state, the net settles into and cycles repeatedly through a recurrent set of about $2^{N/2} = 2^{5000} \approx 10^{1500}$ states.

(2) The net has about $10,000/e \approx 3700$ such recurrent patterns of behaviors. The system must settle into one of these.

(3) A minimal perturbation, defined as reversing the state of a single model gene, is almost certain to move the system from its current pattern of behavior to some other dynamic pattern.

(4) The various possible minimal perturbations to any one pattern can cause the system to jump from that pattern of behavior directly to very many of the other 3700 behaviors.

In sharp contrast, a model genetic net with 10,000 "genes" in which a majority of about 60% or more have one or more canalizing control variables, typically has the following properties:

(1) When released from an initial state, the net settles into and cycles repeatedly in a recurrent pattern through about $\sqrt{N}$ = $\sqrt{10,000}$ = 100 states out of its potential $2^{10,000} \approx 10^{3000}$ states. Behavior is thus enormously restricted.

(2) When released from any initial state, the system must settle into one of about $\sqrt{N}$ = $\sqrt{10,000}$ = 100 such recurrent patterns, each comprised of a distinct set of about 100 states through which the net cycles.

(3) For about 90% of all minimal perturbations, the system returns to the recurrent pattern of behavior from which it was perturbed, exhibiting homeostasis.

(4) The set of all minimal perturbations can cause the net to jump directly from any one recurrent dynamic pattern to only five or six others of the 100 possible behavior patterns. There is a local topology of neighboring behavior patterns.

Enormously restricted, orderly dynamic behavior occurs in virtually any large net of binary genes built using canalizing functions. If the typical large scale dynamic behaviors of this class of systems parallels some large scale behaviors of metazoan cells, those behaviors may be explicable as consequences of these simple local properties.

Typical large scale properties which occur in almost all ensemble members must be insensitve to details of network construction. The large scale properties of cells which are candidates to parallel such average, structurally insensitive, properties of this ensemble of control systems must be those which depend on general features of control systems, not their detailed architecture. These are likely to be properties which are universal, and occur in all or most organisms, whose diversity of detailed control systems precludes common behaviors depending upon those diverse details. The natural biological observables which arise are therefore unlike the small isolatable fragments of cellular systems capable of reasonably complete description which are usually studied. In early attempts to link small scale properties of cell control systems with their large scale dynamic behavior, the suggestion is to consider those large scale properties which depend the least on detailed construction, for there hope of explanation with incomplete knowledge is best.

The following large scale properties of cells seem universal and appear to parallel those of the ensemble of control systems built using canalizing functions of few variables:

(1) The temporal pattern of gene activities corresponding to one cell type in any organism must be enormously restricted in comparison to the potential combinations of gene activities, to a small number of states, or combinations of genes activities through which the cell "modulates" in its ongoing activity. A vast number of possible gene control systems are incapable of this restriction, in particular, large gene control systems built with *non*-canalizing functions. However, temporal gene activity patterns, each with enormous restriction to 100 out of 10^{3000} states for a 10,000 gene net, are expected if canalizing functions are used.

(2) Any organism possesses a particular number of stable, distinct cell types. The number of cell types in an organism increases with the number of its genes and complexity of its DNA. Such a correlation should reflect general features of control organization in all organisms. Estimates of numbers of cell types

and of genes are difficult. Previously (Kauffman, 1969*a*) I presented evidence that the number of cell types in an organism increases as roughly a square root function of the haploid DNA content of its cells, from about one or two cell types for bacteria, to roughly 10^2 for man. Were I to have overestimated the number of genes in higher cells, by not allowing for redundant DNA, by an order of magnitude, it would still appear that the number of cell types increases with the number of genes to a *fractional* power. A stable distinct cell type can be interpreted as a distinct steady or perhaps cyclic pattern of behavior into which an entire cellular control system settles. If the gene system has 100 distinct patterns of gene activity, then each corresponds to the ongoing activity of one cell type; for example, one activity pattern corresponds to cardiac muscle, another distinct pattern to lymphocyte, etc. It is easy to build control systems in which the number of cell types increases as the number of genes to a power greater than 1, or even 2. Constraint to the class of canalizing functions yields an ensemble of control systems in which the number of alternate recurrent behavior patterns, or cell types, increases roughly as a square root function of the number of model genes.

(3) Cell types and model cell types exhibit homeostasis, remaining the same cell type in the face of a wide variety of perturbations. This property does not occur if non-canalizing functions are used.

(4) In virtually all developing systems, no cell type differentiates directly into more than two to five or six other cell types, although it may indirectly develop into many by repeated branching differentiation. Control systems using non-canalizing functions are able to pass from any cell type to a large number of other cell types with minimal perturbations or signals. By contrast, in control systems using canalizing functions, typically one cell type can be induced by small perturbations or signals to differentiate to only a few neighboring cell types, although most perturbations leave it the same cell type.

In asserting that these dynamic properties of randomly constructed model systems parallel those of cells, I am not claiming that control systems evolved over two billion years are random.

Rather, with respect to these particular global dynamic properties, cell control systems may be typical of the ensemble studied by random sampling.

Although the similarities in behavior between control systems using binary switching variables and homologous systems using realistic continuous equations are not straightforward, there are indications (Glass & Kauffman, 1973; Kauffman, 1970; Newman & Rice, 1971) that the highly orderly dynamic behaviors of the ensemble of control systems studied using Boolean canalizing functions also occur when continuous canalizing functions are used.

6. THE DESIGN OF LARGE CONTROL CIRCUITS: EXTENDED FORCING STRUCTURES

The ensemble of control systems built using currently known small scale properties is a conceptual tool which allows both study of the implications of those properties for large scale behavior, and, by examination of average properties of the "wiring diagrams" of ensemble members, yields hypotheses about the most probable kinds of intermediate sized circuits to expect as consequences of the known local properties. Restriction to use of canalizing functions with few controlling variables makes probable the existence of simple and powerful intermediate sized circuits, which I call *extended forcing structures*.

The simplest forcing structure to picture is a familiar model which consists in many hierarchically arranged genes, each controlled by several others and activated if any one of its controlling variables is active. If any gene is active, that alone suffices to activate all the genes it directly controls whether other variables controlling those genes are active or inactive. The active state of a gene propagates directly or indirectly to all members of the hierarchy below it. The inactive state of a gene is not guaranteed to propagate. The hierarchical batteries of genes proposed by

Britten & Davidson (1969, 1971†) are this kind of forcing structure.

Although such positive control cascade derepression circuits have not yet been found, they are members of a more general class of control systems using positive and negative control of which examples are known. The important properties of the cascade derepression hierarchy are: (1) that each gene is regulated according to a canalizing function, in this case each gene is activated if any one gene controlling it is active; (2) the canalized state of each gene, in this case it is the active state which can be determined by one control variable alone, is also the state of that gene which canalizes the genes it in turn directly controls, regardless of the states of other variables controlling those genes. Taken together, these properties assure that the canalizing state, here the active state, propagates in the hierarchy, but the non-canalizing or inactive state may not propagate.

These properties define a transitive relation between two regulated processes like gene activity I term *forcing*: Process A forces process B if: (1) process A is canalized by one or more control variables; (2) process A is a canalizing control variable of process B; (3) the canalized state of A is the state of A which canalizes B. For example, the *lac* promoter has five canalizing control variables which canalize it to $P = 0$. *Lac* promoter is a canalizing control variable of the structural genes, and $P = 0$ is the state of P which determines that transcription fails. Therefore, the *lac* promoter forces the *lac* structural genes. $P = 0$ forces $(ZYA) = 0$.

---

† The control systems suggested by Britten & Davidson (1969) are a specific type of extended forcing structure. They make the interesting suggestion that redundant DNA serves as *cis* regulatory sites for structural genes, and also as templates for synthesis of diffusible regulatory molecules. Repetitive sequences give multiplexing of regulatory variables. All genes are presumed to be non-specifically repressed by histones, and to be activated by any specific input. The hierarchical batteries of genes created are extended forcing structures, for every gene realizes the canalizing OR function on all specific control variables, and the canalized state of each gene —active—canalizes its targets. In their model all specific connections are forcing. This property is false in bacteria and phage. *Lac* operator does not force *lac* structural genes. There is no reason, however, why their concept of hierarchical batteries cannot be widened to the more general notion of extended forcing structures, for which redundant DNA might provide some of the regulatory circuitry.

Two processes canalized by any or all control variables can also be connected so neither forces the other. The *lac* operator is itself canalized by lactose or repressor to the state $Op = 0$. The operator canalizes the structural genes, since $Op = 1$ determines transcription fails. However, the canalized state of Op is 0, not 1. Since the canalized state of Op is not the state which canalizes the structural genes, the operator does not force the structural genes.

By definition, the relation "process A forces process B" is transitive, so that if B also forces C, then A forces C indirectly through B. In this way, extended forcing structures may be constructed. The forcing structures may or may not contain forcing loops.

The following are the major characteristics of extended forcing structures (Fig. 2):

(1) A canalizing state introduced to any element in the structure canalizes all descendent elements, since the canalized state propa-

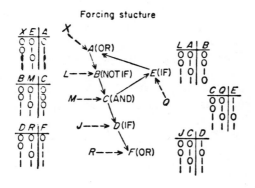

FIG. 2.

gates deterministically to all descendent members regardless of the states of any other control variables. A *non*-canalized state is not guaranteed to propagate. In Fig. 2, if A is in its canalizing state, 1, it determines that B goes to state 0, whatever the state of B's other controlling variable, L. $B = 0$ in turn determines that C goes to 0 whatever the state of M, C's other controlling variable. $C = 0$ determines $D = 1$ and $E = 1$. $E = 1$ forces $A = 1$, completing the forcing loop. If A is in its *non*-canalizing state, 0, it is unable by itself to determine the next state of B. The *non*-canalizing state is not guaranteed to propagate in the structure.

(2) The canalized state may differ at different points in the structure. $A = 1$ forces $B = 0$, etc.

(3) The canalized state of a process is identical for all its canalizing control variables, therefore extended forcing structures can have redundancy of control.

(4) Any forcing loop is a positive feedback loop with a maximum of two steady states. Any loop has a stable steady state when each element is in its canalized state. In that state, the loop is insensitive to all external regulatory events. Some forcing loops also have a metastable steady state with each element in its non-canalized state. In that state, the loop is sensitive to external control variables. In Fig. 2, $A = 1$, $B = 0$, $C = 0$, $E = 1$ is the insensitive forced state. No states of X, L, M, or Q, additional variables controlling loop members, can alter the forced state of the loop; $A = 0$, $B = 1$, $C = 1$, $E = 0$ is the complementary sensitive metastable state, which can be altered by appropriate states of X, L, M, or Q.

(5) If a forcing loop is fixed in its forced steady state, all members of its descendent forcing structure become fixed in their forced states.

(6) Distinct forcing structures may be coupled to one another through a variety of non-forcing control connections allowing subtle and flexible control relations between forcing structures. One forcing structure is sensitive to another only through those of its components which are not in their canalized state.

The power of forcing structures, in which any one gene has one state capable of determining the subsequent states of all descendent members of the structures, are to be contrasted with those obtained if non-canalizing functions are used. By definition, in a non-canalizing function of two or more variables, no single controlling variable alone can determine any state of the target process. To determine one state of the target requires specifying the states of at least two controlling variables; to specify that each of those two is in its correct state requires specifying for each the states of at least two of its controlling variables, and so on. For example, suppose every gene in Fig. 2 realized the non-canalizing *Exclusive Or* function (Table 3). Then no state of any single variable alone could determine any subsequent behavior in the structure. Further, the loop *ABCE* possesses no steady state insensitive to external variables.

These behaviors of forcing structures and non-forcing circuits are independent of the fact that each process in such a structure may itself be a nearly irreversible process. For example, binding of *lac* repressor to the *lac* operator is strong, nearly irreversible. Repressor is removed by the binding of allolactose, which weakens the repressor operator bond. Such loci could be regulated, as is the *lac* operator, according to a canalizing function, or a non-canalizing function like *Exclusive Or*, shown in Table 3.

If all the variables controlling some process derive from the same forcing structure, the regulated process will become fixed in one state if the forcing structure becomes fixed. Forcing structures may be generalized to include such added elements. A forcing loop may use such an added process, whose output feeds back to the forcing structure and canalizes its target. Extended forcing structures with all the properties ascribed to those constructed of idealized Boolean elements, can easily be built using more realistic Michaelis–Menten or sigmoidal binding curves (Kauffman, 1970).

Cellular control systems are likely to contain reasonably large extended forcing structures. Most regulated genes discussed above have from two to six canalizing control variables. It can be shown (Kauffman, 1970, 1971; Glass & Kauffman, 1972) that if the mean number of canalizing variables per controlled process is two or more, then typical members of the ensemble of such systems possess large extended forcing structures.

7. LAMBDA APPEARS TO CONTAIN AN EXTENDED FORCING STRUC' TURE

The occurrence of extended forcing structures may not be purely hypothetical, for one may exist in bacteriophage lambda. Lambda phage has two modes of existence, as a stable lysogen integrated into *E. coli* DNA, and a lytic phase during which the lysogen is excised, replicates, forms mature particles and ruptures the cell. Lambda is not passive while a lysogen. It maintains the lysogenic state by continual synthesis of lambda repressor, the C_1 product, which binds to the left and right operators, O_1 and O_r, and prevents transcription outside the immunity region (Ptashne, 1967; Szybalski *et al.*, 1970). In addition, it appears (Kumar, Calef & Szybalski, 1970) that C_1 maintains its own synthesis by a positive feedback loop in which it facilitates the maintenance promoter, *Prm* (Heinemann & Spiegleman, 1970), or other factors involved in its own synthesis during lysogeny. Lysis is initiated by a signal, presumably due to a metabolic alteration in damaged cells (Ptashne, 1971), which removes repressor from O_1 and O_r and allows waves of leftward and rightward transcription to begin (Szybalski *et al.*, 1970). A sequence of repressive effects due directly and indirectly to gene *tof* sweeps behind the initiated transcription, insuring that the wave be transient. I wish to show this repressive sequence forms a forcing structure.

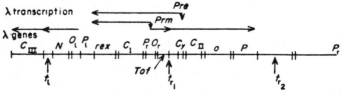

FIG. 3. The regulatory region of λ.

Bacterial polymerase core, sigma factor, and bacterial termination factor Rho (Roberts, 1969) are presumably always present and play no control role in lambda's life cycle. Under this assumption, synthesis of gene *tof* is determined solely by the state of the right operator, O_r (Szybalski *et al.*, 1970). O_r is repressed in the lysogenic state by C_1, but the metabolic signal (X) inducing lysis suffices to remove repressor from O_r. Thus, O_r is governed by the canalizing Not If function, and either absence of C_1 product or presence of X is alone sufficient to determine that O_r is unbound. *Tof*, therefore, is governed by the canalizing If function on C_1 and X; $C_1 = 0$ or $X = 1$, each determines *tof* = 1 (Table 8). In turn, *tof* is a control variable to the maintenance pathway (MP) for C_1 synthesis from the maintenance promoter, *Prm* (Reichardt & Kaiser, 1971). Presuming C_1 facilitates that pathway, the maintenance pathway, MP, is governed by the Not If function, and *tof* = 1 suffices to determine that $MP = 0$ even in the presence of C_1 product (Szybalski *et al.*, 1970; Eisen & Ptashne, 1971). Since *tof* = 1 is the canalized state of *tof* itself, *tof* = 1 *forces MP* = 0.

TABLE 8

C_1	X	O_r	*tof*
0	0	0	1
0	1	0	1
1	0	1	0
1	1	0	1

As described earlier, *tof* is a canalizing control variable for the left operator, O_1. Whatever the states of C_1 and X, the other variables controlling O_1, *tof* = 1 determines that O_1 is bound. Since *tof* = 1 is the canalized state of *tof*, *tof* = 1 *forces* $O_1 = 1$.

By binding to O_1, *tof* prevents synthesis on the left operon, notably gene N, which is required for adequate transcription beyond t_1, the termination site on the left operon, and t_{r1} and t_{r2} on the right operon (Szybalski *et al.*, 1970; Thomas, 1971). Blocking O_1 prevents C_{III} synthesis both directly since C_{III} is transcribed from P_1, and indirectly, since it is beyond t_1 and requires N, probably for antitermination. Since N is required to transcribe beyond t_{r1} and t_{r2}, $N = 0$ blocks (almost completely) synthesis of

C_{II}, which is distal to t_{r1} and Q, which is distal to t_{r2}. Q is necessary for late right transcription (Echols, 1971). C_{III} and C_{II} are jointly required for synthesis of C_1 by the establishment pathway (*EP*) from the establishment promoter *Pre* (Reichardt & Kaiser, 1971) during lysogenization. Therefore, *tof* = 1 forces, through O_1 and N, $C_{II} = 0$, $Q = 0$ and $EP = 0$. The forcing structure from *tof* is shown in Fig. 4.

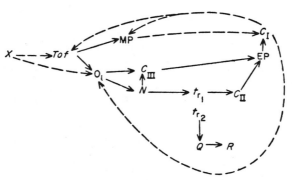

FIG. 4. The forcing structure in lambda. Solid arrows are forcing connections; dotted arrows are non-forcing connections.

C_1 has two pathways, establishment and maintenance, for its synthesis; either alone suffices. *Tof* = 1 insures that *MP* = 0 and *EP* = 0. While *tof* = 1 does not force $C_1 = 0$ on either *MP* or *EP* alone, both pathways are members of the forcing structure headed by *tof* and *MP* = 0 and *EP* = 0 jointly determine $C_1 = 0$. C_1 may therefore be added as a member of the forcing structure headed by *tof*.

In turn, C_1 is a control variable for *tof*; $C_1 = 0$ determines *tof* = 1 (Szybalski *et al.*, 1970; Eisen & Ptashne, 1971). *Therefore, the C_1 tof loop is a forcing loop with two steady states*; $C_1 = 1$, *tof* = 0 is the metastable steady state corresponding to lysogeny. Its metastability is shown by the response to the inducing signal, X, which derepresses O_1 and O_r, initiates lysis, and places *tof* in its canalizing state, 1. The stable forced steady state is $C_1 = 0$, *tof* = 1. The waves of transient transcription and translation following induction from the metastable lysogenic state precede the propagating inhibitory canalizing effects of $tof = 1$, and comprise the lytic sequence during which *E. coli* is destroyed. Therefore, the predicted stable steady state, $C_1 = 0$, $tof = 1$, is not normally observed. However, it can be observed in mutants (Neubauer & Calef, 1970) which interrupt the lytic sequence but leave the C_1 *tof* loop intact. The state $C_1 = 0$, $tof = 1$ corresponds to the non-immune (im^-) state. These mutants clearly exhibit the bistable character of the *tof* C_1 loop, for infected cells are capable of two phenotypes: the im^+ state corresponding to normal lysogeny with $C_1 = 1$, $tof = 0$; and the im^- state. Each regulatory phenotype is transmissible through an unlimited number of cell generations. In addition, a spontaneous conversion from im^+ to im^-, or *vice versa*, occurs at low frequencies. Analysis of the C_1 *tof* circuit as a forcing loop in which the immune lysogen is the metastable state and im^- is the stable forced state with respect to external control variables of the loop, accords with the observation that several environmental stimuli increase the conversion of im^+ to im^-, while no environmental situation specifically converting im^- to im^+ has been found (*ibid.*).

8. POTENTIAL FUNCTIONS OF EXTENDED FORCING STRUCTURES

Lambda possesses the most complex control circuitry yet analyzed in detail. Despite differences between phage and higher cells, the property of canalization depends only on the properties of monotonicity and saturation in the binding curves of ligands controlling activity of biosynthetic elements. Since Langmuir and sigmoidal binding curves are common, canalization and forcing structures potentially link processes at many distances from direct gene activity. In this sense, lambda provides a model for the kinds of organization of regulated processes which might be expected in higher cells. Lambda possesses an extended forcing structure with a forcing loop. The former allows a single gene to control directly or indirectly the activities of many other genes. The latter gives lambda the capacity for heritable differentiation into two regulatory states, im^+ and im^-. Both properties would clearly be useful in higher cells.

9. EXPERIMENTAL TESTS

The ensemble of control systems constructed using currently known local properties is a conceptual device with at least two related uses. (1) It is a tool to assess the implications of those local properties for probable large scale dynamic behaviors and circuitry. (2) If the typical large scale properties of the ensemble fit cells, cells are probably typical members of the ensemble with respect to those properties. Those properties must therefore be insensitive to details of construction, and must reflect general features of the control organization which are consequences of the local properties generating the ensemble. *Thus, the ensemble approach suggests that a number of large scale cellular properties are to be explained as consequences of small scale features of the organization which are present everywhere in the control system, rather than through discovery of additional specific parts which cause the large scale features.* If so, the appropriate experimental work consists in establishing the small scale properties, and the correspondence between predicted and observed large scale properties. It is an inescapable feature of an ensemble approach that it makes no prediction about the detailed design of any particular circuit, since it is a statistical theory which predicts the *kinds* of circuits and large scale dynamics to expect.

Of the four large scale dynamic behaviors discussed above, it seems clear that any cell type is an enormously restricted pattern of gene activity, exhibits homeostatic return to the same cell type after a variety of perturbations, and can directly differentiate into only a few other cell types. The deduction that cell types should increase as roughly a square root function, or a fractional power, of the number of genes may prove false. If so, the ensemble studied is too broad, and additional features of the organization will have to be found to refine the ensemble. In the present ensemble, there tends to be one very large extended forcing structure in any control system (Kauffman, 1970, 1971). Were there instead a number of distinct forcing structures, the general effect would be to increase the number of steady behaviors, the cell types, while leaving the other global properties unaltered.

The present ensemble makes further predictions about large

scale dynamic properties of cells. Comparison of the patterns of gene activity between different model cell types indicates that on the average, two model cell types have identical activities in about 0·9 of the model genes (Kauffman, 1969a). Data for organisms is sparse. Berendes (1965) compared puffing patterns of 110 loci on giant chromosomes of *Drosophila hydei* after ecdysone treatment in three tissues, malpighian tubules, salivary glands, and stomach. Puffing patterns were identical in about 0·9 of the loci for each of the three possible pairs.

If, as predicted from the ensemble, forcing structures exist in higher cells, their functional independence from other parts of the control system while propagating a canalized state recommends then as natural subsystems into which the cell control system fragments. There appear to be strong candidates for extended forcing structures in higher metazoans. One property of forcing structures is that any single member has one state capable of determining directly or indirectly the states of all descendent members of the structure. This property does not occur readily in circuits where each locus is regulated by more than one variable according to a non-canalizing function. Ohno's (1971) analysis of the *Tfm* mutant in mouse has shown that mutation of this single locus prevents triggering an entire program of development of male secondary sex organs. The sequence of over 100 puff alterations in polytene chromosomes of *Drosophila melanogaster* and related species, induced by the hormone ecdysone (Clever, 1966; Ashburner, 1971), might be extended forcing structures. Britten & Davidson (1969, 1971) constructed hierarchical batteries of genes to account for several aspects of metazoan differentiation where one molecular species controls many unlinked loci. Their hypothetical circuits are restricted types of forcing structures, which require generalization to fit the known case of lambda.†

Forcing loops are circuits having two alternate steady regulatory states and are therefore well suited to underlie the sequence of commitments to alternate regulatory states in development. If so, they ought to occur in *Drosophila* imaginal disks. These disks are groups of cells which become determined early in embryogenesis

---

†See footnote on page 252.

to develop into particular adult structures, wing, leg, antenna, etc., during metamorphosis (Hadorn, 1966; Gehring, Mindek & Hadorn, 1968; Tobler, 1966; Gehring, 1967). In another publication (Kauffman, 1973) I give grounds to think that determination in *Drosophila* is underlain by at least four control circuits, each with two steady states.

Even were that model correct, the bistable circuits might not be forcing loops. Diagnostic properties of forcing loops predict specific classes of mutants. If any member of a forcing loop is mutated to remain constitutively in its canalized state, all other members of the loop and the entire descendent structure should be forced as well. It should not matter which member of the loop is so mutated; if the loop contains more than one member, an equivalence class of mutants with approximately identical phenotypes should exist, each associated with the same co-ordinated abnormality in patterns of gene activity in all tissues. This block should not be able to be bypassed by *normal* control variables to descendent members of the forcing structure. Were the gene mutated to remain constitutively in its *non*-canalized state, the block should be able to be bypassed.

10. CONCLUSION

There are only two ways we shall come to explain large scale properties of cells. Either all properties of interest will be consequences of circuitry sufficiently simple for detailed characterization, or we shall have to use incomplete knowledge and provide statistical explanations. The ensembles of control systems constructed using any currently known small scale properties of the organization should prove useful tools both to predict the organization of large circuits, and to explain large scale behaviors in the face of incomplete knowledge.

Although the use of ensembles of systems and their average properties has been successful in statistical mechanics, the procedures I am suggesting are less familiar in cell biology. The notion that a large scale property of cell control systems (such as the number of cell types into which any one cell type can directly differentiate) is an observable deserving explanation seems premature, precisely because it is probably not explicable from the properties of any small number of genes or processes. Further, the explanation offered seems uncertain. To assert that any single large scale behavior is a consequence of cells being typical members of some ensemble of systems may be false, since cells might be atypical members of the ensemble and the large scale property might be due to other mechanisms. However, reasonably strong evidence that cells are typical members of an ensemble would be provided if many distinct large scale properties of cells simultaneously paralleled typical properties of ensemble members. The observed restriction of regulated genes and processes to canalizing functions of few controlling variables is an excessively simple small scale property which already may predict simultaneously four or five large scale dynamic behaviors of cells, as well as the existence of extended forcing structures. Extensions of this early theory consist in discovering additional large scale properties of cells which are insensitive to details of construction and additional small scale properties through which to explain them. The possibility of an adequate theory of large scale properties of cells as local properties are better understood seems good, even if it should prove impossible to characterize control systems completely.

This work was supported in part by the Alfred P. Sloan Foundation.

REFERENCES

1. M. Ashburner, *Nature, New Biology*, **230** (1971), 222.
2. *The Bacteriophage Lambda* (A. D. Hershey, ed.), Cold Spring Harbor Laboratory, 1971.
3. H. D. Berendes, *Devl. Biol.*, **11** (1965), 371.
4. M. Brenner and B. N. Ames, *Metabolic Pathways*, vol. 5 (H. J. Vogel, ed.), Academic Press, New York, 1971, 349.
5. M. S. Bretscher, *Nature, Lond.*, **217** (1967), 509.
6. R. J. Britten and E. H. Davidson, *Science, New York*, **165** (1969), 349.
7. ———, *J. Theor. Biol.* **32** (1971), 123–130.
8. C. Burstein, M. Cohn, A. Kepes, and J. Monod, *Biochem. Biophys. Acta.*, **95** (1965), 634.

9. U. Clever, *Devl. Biol.*, **14** (1966), 421.

10. B. De Crombrugghe, B. Chen, W. Anderson, P. Nissley, M. Gottesman, and I. Pastan, *Nature, New Biology,* **231** (1971), 139.

11. H. Echols, *The Bacteriophage Lambda* (A. D. Hershey, ed.), Cold Spring Harbor Laboratory, 1971, 247.

12. H. Eisen and M. Ptashne, *The Bacteriophage Lambda* (A. D. Hershey, ed.), Cold Spring Harbor Laboatory, 1971, 239.

13. W. Gehring, *Devl. Biol.,* **16** (1967), 438.

14. W. Gehring, G. Mindek, and E. Hadorn, *J. Embryol. Morph.,* **20** (1968), 307.

15. L. Glass and S. A. Kauffman, *J. Theor. Biol.*, **34** (1972), 219.

16. ———, *J. Theor. Biol.,* **39** (1973), 103.

17. E. Hadorn, *Devl. Biol.*, **31** (1966), 424.

18. S. F. Heinemann and W. G. Spiegleman, *Proc. Nat. Acad. Sci. U.S.A*, **67** (1970), 1122.

19. F. Jacob and J. Monod, *21st Symp. Soc. Study of Development and Growth,* Academic Press, London, 1963.

20. S. A. Kauffman, *J. Theor. Biol.*, **22** (1969a), 437.

21. ———, *Nature, Lond.,* **224** (1969b), 177.

22. ———, *Mathematics in the Life Sciences*, Amer. Math. Soc., **3** (1970), 63.

23. ———, *Current Topics in Development Biology*, Academic Press, New York, **6** (1971), 145.

24. ———, *Science, New York,* **181** (1973), 310.

25. S. Kumar, E. Calef, and W. Szybalski, *Cold Spring Harbor Symp. Quant. Biol.*, **35** (1970), 331.

26. B. Muller-Hill, H. B. Rikenberg, and K. Wallenfels, *J. Molec. Biol.* **10** (1964), 303.

27. S. Newman and S. Rice, *Proc. Nat. Acad. Sci. U.S.A.*, **68** (1971), 92.

28. Z. Neubauer and E. Calef, *J. Molec. Biol.*, **51** (1970), 1.

29. S. Ohno, *Nature, Lond.,* **234** (1971), 134.

30. M. Ptashne, *Nature, Lond.,* **214** (1967), 232.

31. ———, *The Bacteriophage Lambda* (A. D. Hershey, ed.), Cold Spring Harbor Laboratory, 1971, 221.

32. L. Reichardt and A. D. Kaiser, *Proc. Nat. Acad. Sci. U.S.A.*, **68** (1971), 2185.

33. A. D. Riggs, R. F. Newby, and S. J. Bourgeois, *J. Molec. Biol.*, **51** (1970), 303.

34. J. W. Roberts, *Nature, Lond.*, **224** (1969), 1168.

35. W. Szybalski, K. Bovre, M. Fiandt, S. Hayes, Z. Hradecna, S. Kumar, H. A. Lozeron, H. J. J. Nijkamp, and W. F. Stevens, *Cold Spring Harbor Symp. Quant. Biol.,* **35** (1970), 341.

36. R. Thomas, *The Bacteriophage Lambda* (A. D. Hershey, ed.), Cold Spring Harbor Laboratory, 1971, 211.

37. H. Tobler, *J. Embryol. Exp. Morph.,* **16** (1966), 609.

38. H. E. Umbarger, *A. Rev. Biochem.*, **38** (1969), 323.

39. H. J. Vogel, *Metabolic Pathways,* vol. 5, Academic Press, New York, 1971.

40. R. Vogel, W. McLellan, A. Hiroven, and H. Vogel, *Metabolic Pathways*, vol. 5, (H. J. Vogel, ed.), Academic Press, New York, 1971, 463.

41. G. Zubay and D. A. Chambers, *Metabolic Pathways,* vol. 5 (H. J. Vogel, ed.), Academic Press, New York, 1971.

42. G. Zubay, L Gielow, and E. Englesberg, *Nature, New Biology,* **233** (1971), 164.

PATTERNS OF PHASE COMPROMISE IN BIOLOGICAL CYCLES

A. T. Winfree

INTRODUCTION

INTRODUCTION

One way to capture the time-independent essentials of a complex dynamical process is to describe the topological relations in its state space among whatever special manifolds its state approaches in the limits of large positive or negative time. To take an example from cell biology, Fitzhugh's reduction of Hodgkin and Huxley's ordinary differential equation organizes the description of nerve membrane behavior by exhibiting (as a function of ionic parameters) the geometrical relationships among one or three generic stationary states, a separatix, and/or a limit-cycle trajectory in the four-dimensional state space defined by membrane potential, sodium and potassium activations, and sodium inactivation (Fitzhugh, 1960; Zeeman, 1972).

One way to explore the state space of a dynamical system is to systematically perturb it, seeking to map the locations of stationary states, separatrices (thresholds), limit cycles, etc. (e.g., Winfree (1970, 1972), on circadian rhythms and on glycolysis).

These critical states and other special loci show up as

boundaries separating regions of qualitatively different system response, loci of time-translated but otherwise identical responses, etc., in a space describing the stimulus (e.g., its time of application, duration, and/or intensity) and as isolated singularities. Each stimulus puts the system in some new state; the *set* of stimuli provides a *set* of new states, a locus of dimension not exceeding the number of independently-varied stimulus parameters. It is this locus of states whose properties we explore by systematic perturbation: from these states as initial conditions we observe the system's asymptotic behavior, partitioning the stimulus space into domains of qualitatively different system response. The geometry of these domains reflects something about the geometry of special loci in state space.

CHEMICAL MIXING

But unless enough is known about the dynamics of the perturbed system (i.e., during the stimulus), the mapping from stimulus space into state space (thus the geometry of the set of accessible initial conditions) may be too much in doubt to permit strong inference about the *un*perturbed system's dynamical organization. In this dilemma, a particularly simple and natural perturbation, though applicable only to periodic biochemical systems, recommends itself for the explicitness of the map it induces from stimulus parameter space into the state space (which, for biochemical systems, is a chemical concentration space).

Given a stable, autonomous biochemical oscillation distributed homogeneously in solution or suspension, one can physically mix two portions at different stages of their cycle, in various proportions by volume. Unless semipermeable membranes intervene, all reactant concentrations are instantly adjusted by mixing, to a compromise in exact proportion to the volume ratio. It is thus possible to explore and exactly map the contents of a region of the concentration space (which we might call the "innards" of the cycle), viz., the set of all states on straight lines connecting states on the cycle. This set has at most three dimensions, corresponding to the volume ratio and the phases of the volumes mixed; if the cycle lies wholly in a 3- or 2-dimensional section through concentration space, then the map from this 3-dimensional parameter space into concentration space is partly redundant. This can be helpful in checking experimental results for reproducibility and consistency.

Experiments of this sort have been undertaken in three areas of cell biology: (1) Rusch, Sachsenmaier, Behrens and Gruter (1966), Guttes and Guttes (1969), Guttes, Devi and Guttes (1969), Sachsenmaier, Remy and Plattner-Schoble (1972) and Kauffman and Wille (1975) have explored control of the cell cycle by fusing plasmodia of synchronously-mitosing *Physarum polycephalum*, and recording the compromise phase of synchronous mitosis established within the next ten hours. Rao and Johnson (1970) have studied rephasing of the cell cycle by fusion of mammalian cells; Dewey, Miller,and Nagasawa (1973) have reported a mitosis-synchronizing interaction between hamster ovary cells mixed in tissue culture; (2) Brinkmann (1966, 1967) and Edmunds (1971) have inquired whether *Euglena* cells synchronize one another's circadian rhythms through diffusible metabolites, by mixing suspensions with $\frac{1}{4}$ to $\frac{1}{2}$ cycle phase difference, then comparing the resultant motility rhythms with the superposition of the two unmixed control rhythms. Hastings and Sweeney (1957) reported similar experiments using circadian rhythms in another protozoan population; (3) Ghosh, Chance, and Pye (1971) have exhibited interactions among the glycolytic oscillations of yeast cells by exploring the pattern of phase compromise between two equal aliquots of cell suspension, mixed at various time in both cycles.

In all these cases, the mixing is incomplete, since not all the pertinent molecular species are free to diffuse through the nuclear envelope and/or plasma membrane. Yet these experiments, and their results, share certain symmetries with each other and with the idealized mixing experiment treated in more detail below. My purpose here is to explore theoretically some of the consequences of these symmetries.

THE GENERAL ARGUMENT

Let us first establish a convenient notation: Let phase at a prescribed moment, $\phi \in S^1$, be defined as the fraction of a cycle elapsed since the most recent occurrence of a marker event, e.g., mitosis in *Physarum*, or the peak of NADH fluorescence in yeast cells. Let a volume v, expressed as a fraction of the total volume of the mixture, be mixed at phase ϕ_1 with volume $1 - v$ at phase ϕ_2. Let $\phi_3 \in S^1$ denote the phase of the *mixture* (which I will sometimes call the "resultant") measured later but extrapolated back to the moment of mixing or measured an integer number of cycletimes after mixing. This definition presupposes that the mixture eventually continues to oscillate with the same period and "waveform" as either control aliquot, differing only in phase, and that the resulting phase ϕ_3 is *uniquely* defined in S^1. If for some range of $\phi_1 \times \phi_2$, there are two or more alternative ϕ_3 results, then the inferences discussed below cannot be rigorously drawn.

Though phase is defined on the circle rather than on the real number axis, it will be convenient to implicitly map the circle repeatedly onto the real axis: once from 0 to 1, around again from 1 to 2, and so forth. Thus phase can be described by a *number* (modulo 1), for which arithmetic addition, subtraction, etc., are familiar operations. This also makes it possible to sketch phase functions on unit squares and cubes that would otherwise have to be drawn on tori and hypertori.

The resulting ϕ_3 may be expressed as a single-valued function R (for "resultant") of ϕ_1, ϕ_2, and v with these properties:

(A) If the oscillation underlying the observed rhythm is in all respects periodic with unit period, then R is periodic with unit period in ϕ_1 and ϕ_2: $\phi_3 = R(\phi_1, \phi_2, v) = R(\phi_1 + n, \phi_2 + m, v)$ for any integers, m, n. In other words, only the volume ratio and the two phases determine the resultant, independent of the cycle in which either phase is taken.

(B) R is invariant under exchange of labels: $\phi_3 = R(\phi_1, \phi_2, v)$ $= R(\phi_2, \phi_1, 1 - v)$. In other words, the result doesn't depend on which aliquot we choose to call "1".

(C) Supposing that mechanical mixing and/or fusion has no effect on the oscillation, mixing of synchronous reactions should affect neither: $\phi_3 = R(\phi, \phi, v) = \phi$. In the case of fusing cells or plasmodia, this condition presupposes that only intensive variables (concentrations, temperature, etc.) determine the course of the reactions, and extensive variables (nuclear surface/volume ratio, total mitochondrial biomass per fused cell, etc.) are irrelevant.

It should be noted that none of the above require chemical interaction between the mixed volumes: averaging of sinusoidally varying concentrations without physical contact also satisfies (A), (B), and (C).

These features have a surprising consequence, which I think is important to distinguish explicitly as a purely logical consequence of the definition of "phase" regardless of the mechanism underlying the oscillation: the result *cannot* depend in a continuous way on ϕ_1, ϕ_2, and v; there must exist a locus in $\phi_1 \times \phi_2 \times v$ space along which the resulting phase is indeterminate. The *way* in which this happens, i.e., the geometry of the locus, may betray something about the underlying mechanism.

To see why this locus of indeterminacy or discontinuity is necessary, consider Fig. 1. The horizontal and vertical axes are the phases ϕ_1 and ϕ_2 at the moment of mixing, plotted repetitiously through two full cycles in each direction. Topologically, $\phi_1 \times \phi_2$ is a torus, here slit open along both generating circles and flattened onto the plane. The resultant ϕ_3 may be imagined plotted upward, out of the paper as a periodically multiple-valued surface above

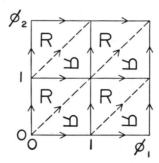

FIG. 1. The $\phi_1 \times \phi_2$ torus is depicted in the plane, twice along each phase axis. In each of these four unit cells the dependence of ϕ_3 on ϕ_1 and ϕ_2 is the same, by assumption A. The ϕ_3 surface could be portrayed by ϕ_3 level contours in the $\phi_1 \times \phi_2$ plane; here the letters R replace this contour map in each unit cell. By assumption B, ϕ_3 depends equally and symmetrically on ϕ_1 and ϕ_2 at $v = \frac{1}{2}$, so the contour map R is symmetric about the main diagonal in each unit cell. Arrows indicate the direction of phase increase along the boundary of each triangle. (The dashed lines represent mirror planes, not contour lines, of ϕ_3; they are perpendicular to local ϕ_3 contours.)

the $\phi_1 \times \phi_2$ plane, or as cartographers do, as a contour map of that surface drawn on the plane. In either case, $\phi_3 (\phi_1, \phi_2)$ is denoted simply by the symbol R, which in Fig. 1 exhibits features A and B at $v = \frac{1}{2}$. Fig. 2 sets Fig. 1 into the $\phi_1 \times \phi_2 \times v$ cube at $v = \frac{1}{2}$ and illustrates feature (C) on the diagonal plane.

Since by (B), ϕ_1 and ϕ_2 are interchangeable at $v = \frac{1}{2}$, the main diagonal (dashed) is a mirror symmetry axis. Because of this and because ϕ_3 is periodic along both axes (A), ϕ_3 must change through the same integer number k of full cycles along any one of the perpendicular legs of the small triangles in Fig. 1 in the indicated directions. According to (C), ϕ_3 changes continuously and uniformly through one cycle along the increasing diagonal. Thus in circumnavigating any one of the small triangles, ϕ_3 rises or falls through $J = k + k - 1$ cycles. If ϕ_3 is continuous then the same must be true on any distortion of this closed loop path, for an integer cannot change gradually, nor can it switch abruptly as the loop is gradually deformed unless ϕ_3 is somewhere discontinuous or undefined. So J is an odd integer on any closed path. But in circumnavigating any full square, ϕ_3 changes through $J' = k + k - k - k = 0$ full cycles; the same is true on any sufficiently small closed path, supposing only that ϕ_3 is continuous and defined everywhere. Zero is not odd. That's odd.

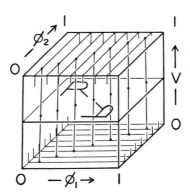

FIG. 2. Fig. 1 is set into context as the $v = \frac{1}{2}$ plane in a $\phi_1 \times \phi_2 \times v$ cube. Along the diagonal plane (indicated by vertical contour lines with 3 dots), assumption (C) says $\phi_1 = \phi_2 = \phi_3$. Additionally, the result (ϕ_3) is independent of ϕ_1 at $v = 0$ and independent of ϕ_2 at $v = 1$: $R (\phi_1, \phi_2, 0) \equiv \phi_2$ and $R (\phi_1, \phi_2, 1) \equiv \phi_1$.

R. Casten drew to my attention a similar argument based on Fig. 2. Consider any square at fixed ϕ_1. As (ϕ_2, v) circumnavigates it, ϕ_3 passes through one cycle: there is no change across the top (at $v = 1$, ϕ_2 doesn't matter), some change $\Delta\phi$ going down the front side (at $\phi_2 = 0$), one cycle across the bottom ($\phi_3 = \phi_2$ at $v = 0$), and a change $-\Delta\phi$ going up the back side (at $\phi_2 = 1$). Thus ϕ_3 is undefined or discontinuous somewhere within each square: the phaseless set is not only a pair of isolated points at $v = \frac{1}{2}$, but is a *curve* penetrating the cube.

Let us now examine the shape of this phaseless locus under two different hypotheses about the underlying periodic mechanism: that it is the simplest kind of smooth limit cycle, and that it is the simplest kind of relaxation oscillator. Though these examples are simplistic and unrealistic, I believe their qualitative, topological features will be retained in the more complex models one might employ in modelling glycolytic oscillations in yeast cells or the mitotic cycle in *Physarum*, for example. My intent in describing idealized cases is to provide a framework within to examine the idiosyncrasies of less "ideal" experimental material.

MIXING ON LIMIT CYCLES

Imagine a circular limit cycle involving only two[1] important molecular species and defined in polar coordinates about $(1, 1)$ by $\rho = r(\phi)$. In cartesian coordinates denote the concentrations $x \geqslant 0$ and $y \geqslant 0$. The changing state (x, y) of each reaction will be represented by a moving dot in the first quadrant of the (x, y) plane (Fig. 3). Let volumes v and $(1 - v)$ be mixed at states (x_1, y_1) and (x_2, y_2) respectively. The pooled concentrations immediately after rapid mixing are:

$$x_3 = vx_1 + (1 - v)x_2,$$
$$y_3 = vy_1 + (1 - v)y_2, \tag{1}$$

---

[1] Hanusse (1972) and Tyson and Light (1974) show that to realize a repelling stationary state in an open biochemical system with bimolecular kinetics requires a third reactant. The reader may imagine a third concentration axis on Figs. 3, 4, 5, 10, 16, and 17 without altering the conclusions reached here.

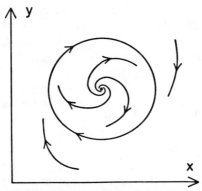

FIG. 3. The arrows indicate rates of change of two chemical concentrations x and y in a dynamical system possessing a unique stationary state and a symmetric limit cycle whose basin of attraction is the whole space minus the stationary state.

as indicated in Fig. 4. What do these quantities mean in terms of phase? Because the system dot is not now on the limit cycle, phase is not immediately defined. It proves convenient to adopt the notion of "latent phase" (Winfree (1967, 1972)) as follows: after mixing we revisit the reaction vessel at unit intervals (the limit-cycle period), when the parent vessels are each time again at

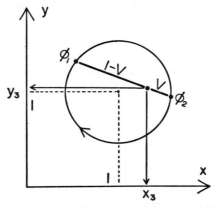

FIG. 4. At phases ϕ_1 and ϕ_2 on the limit cycle, volumes v and $(1 - v)$ respectively are combined as unit volume with concentrations x_3 and y_3 determined barycentrically (text equation 1), i.e., at the center of mass.

(x_1, y_1) and (x_2, y_2) on the limit cycle. If the hybrid mixture is initially within the limit cycle's attractor basin, then it will be seen approaching the limit cycle through a discrete sequence of states eventually clustering at some phase ϕ_3 on the limit cycle. The latent phase Φ is defined as a function of state $\Phi \colon R^2 \to S^1$, by $\Phi(x_3, y_3) = \phi_3$. Every state within the attractor basin has some latent phase. Along every trajectory leading to the limit cycle, latent phase increases constantly: $d\Phi/dt = 1$ because Φ is defined as "ϕ near the limit cycle n cycles later (n large)", and phase ϕ is so defined that $d\phi/dt = 1$. So Φ returns to the same value at unit intervals, as the dot swings closer and closer to the limit cycle. Each locus $\Phi(x, y) = $ constant is a curve which crosses the limit cycle at only one point and terminates on the attractor basin's boundaries (in this case, the stationary state and infinity). These curves (or $(N - 1)$-dimensional surfaces, if the limit cycle is in R^N, $N > 2$) are called isochrons. They collectively foliate the limit cycle's attractor basin and cannot cross since Φ is unique except on the basin's boundaries (see Appendix). Let us suppose the isochrons are perfectly radial equispaced rays:

$$\operatorname{tn} \Phi(x, y) = (y - 1)/(x - 1) \tag{2}$$

(in which I use the notation $\operatorname{tn} z \equiv \tan 2\pi z$, and similarly below). For example, this is nearly the case in the van der Pol oscillation at low-μ limit. On the limit cycle

$$\begin{aligned} x_1 &= 1 + r(\phi_1) \operatorname{cs} \phi_1, \\ y_1 &= 1 + r(\phi_1) \operatorname{sn} \phi_1, \end{aligned} \tag{3}$$

and similarly for oscillator 2. Plugging (3) into (1) and (1) into (2) completes the algebraic derivation of phase mixing of such limit cycle reactions:

$$\operatorname{tn} \phi_3 = \frac{v r(\phi_1) \operatorname{sn} \phi_1 + (1 - v) r(\phi_2) \operatorname{sn} \phi_2}{v r(\phi_1) \operatorname{cs} \phi_1 + (1 - v) r(\phi_2) \operatorname{cs} \phi_2}. \tag{4}$$

For simplicity, in all the above, let $r(\phi) = \frac{1}{2}$, i.e., make the limit cycle concentric to the stationary state, as in Fig. 3, so

$$\text{tn } \phi_3 = \frac{v \text{ sn } \phi_1 + (1 - v) \text{ sn } \phi_2}{v \text{ cs } \phi_1 + (1 - v) \text{ cs } \phi_2}. \tag{5}$$

This equation possesses symmetries (A), (B), (C) above, and additionally deals explicitly with the implicit phaseless locus. Fig. 4 depicts its physical origin: when equal aliquots $\frac{1}{2}$ cycle apart are mixed, (x_3, y_3) is the stationary state $(1, 1)$ at which Φ is undefined, and near which all values of Φ are accessible (Fig. 5). From equation (5), setting $\text{tn } \phi_3 = \frac{0}{0}$, for ambiguous ϕ_3, we find that v must be $\frac{1}{2}$ and $\phi_1 = \phi_2 + \frac{1}{2}$: the phaseless locus is a straight line diagonally penetrating the cube. In a cross-section at fixed $\phi_2 = 0$, equation (5) becomes

$$\text{tn } \phi_3 = \frac{\text{sn } \phi_1}{\text{cs } \phi_1 + (1 - v)/v}, \tag{6}$$

which is sketched in Fig. 6 as a ϕ_3-contour map. At fixed $\phi_2 \neq 0$,

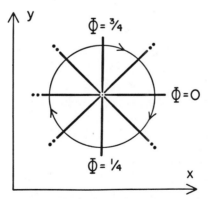

FIG. 5. The coordinate plane of Figs. 3 and 4 is repeated once again to show the "isochrons" or loci of "latent phase" $\Phi(x, y) = $ constant. All mixtures initially on the same isochron, ϕ_3, wind up on the limit cycle synchronously: a large integer number of cycle times after mixing, all such volumes are indistinguishable from a control $\Phi = \phi_3$ which started at the intersection of isochron ϕ_3 with the limit cycle. Isochrons need not be as symmetric as here caricatured, but must cross the limit cycle once only, and converge to the stationary state without crossing each other.

276   *A. T. Winfree*

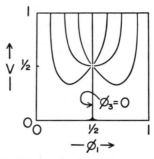

Fig. 6. At $\phi_2 = 0$, the resulting ϕ_3 is shown by contour lines (isochrons) as a function of ϕ_1 and v from equation (6): at small v, $\phi_3 \sim 0$; at large v, $\phi_3 \sim \phi_1$; at $v = \frac{1}{2}$ and $\phi_1 = \frac{1}{2}$, the stationary state is reached.

the shape is the same with coordinate ϕ_1 changed to $(\phi_1 - \phi_2)$ and ϕ_3 changed to $(\phi_3 - \phi_2)$. Fig. 2 is now completed by installing Fig. 6 as the side panel in four positions according to rules (A) and (B) (Figs. 7 and 8). It is apparent that all isochron surfaces converge radially along the phaseless locus.

In Fig. 9 isochron surface $\phi_3 = 0$ is sketched according to equation (5). It resembles a screw dislocation, being bounded on the inside by the phaseless locus and on the outside by a helix composed of line segments crossing each of 4 cube faces in sequence. This screw surface joins smoothly onto its replicas in the four adjacent unit cells.

All other isochrons, $\phi_3 = $ constant, have identical shape, differing from $\phi_3 = 0$ only in a rigid displacement along the phaseless locus.

With a linear coordinate transformation, equation (5) can be written more compactly. Let $a = \phi_3 - \frac{1}{2}(\phi_1 + \phi_2)$, $b = \frac{1}{2}(\phi_1 - \phi_2)$, and $c = 2v - 1$. Then

$$\frac{\text{tn } a}{\text{tn } b} = c \tag{7}$$

as may be verified by direct substitution into equation (5) of ϕ_1 and ϕ_2 in terms of a, b, and c. It is essential to take care about ambiguities of $\pm \frac{1}{2}$ cycle involving the tangent's argument in (7).

Now let us deform the overly symmetric model of equation (5)

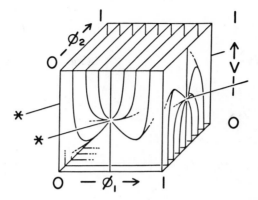

FIG. 7. Fig. 6 (and its inverse for $\phi_1 = 0$ according to assumption B) are added to Fig. 2 as the $\phi_2 = 0$, $\phi_1 = 1$, $\phi_2 = 0$, and $\phi_2 = 1$ planes. The phaseless point becomes a locus ($v = \frac{1}{2}$, $\phi_1 - \phi_2 = \frac{1}{2}$) in the $\phi_1 \times \phi_2 \times v$ cube.

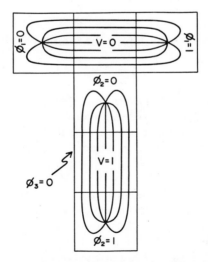

FIG. 8. Contours of constant ϕ_3 on the surface of the $\phi_1 \times \phi_2 \times v$ cube of Figs. 2 and 7, opened flat, show where each isochron and the phaseless locus encounter the cube's faces. ϕ_3 increases from 0 to 1 to the right on face $v = 1$, and from 0 to 1 upward on face $v = 0$. The entire boundary of this figure is at $\phi_3 = 0$.

278   A. T. Winfree

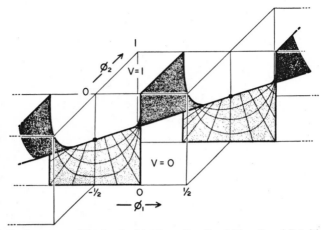

Fig. 9. Isochron $\phi_3 = 0$ is sketched in the unit cells of Figs. 2 and 7 (with the ϕ_1 axis shifted $\frac{1}{2}$ cycle for clarity of view). It connects the phaseless locus to a spiral locus on the cube's surface, the boundary of Fig. 8, and connects onto the duplicate isochron surfaces in 4 adjacent unit cells along the 4 upright elements. All other isochrons are parallel translations of this along the phaseless locus.

and Figs. 3–9, by rigidly shifting the limit cycle off center, so

$$r(\phi) = \tfrac{1}{2} - \tfrac{1}{4} \text{ cs } \phi \qquad (8)$$

in equations (3) and (4), rather than $r(\phi) = \tfrac{1}{2}$ (Fig. 10). Then the phaseless locus is made to bend sinusoidally, arching above and diving below the midplane $v = \tfrac{1}{2}$, with the consequence that each horizontal section through the middle of the cube intersects this isochron convergence locus at *two points*, one on each side of the main diagonal.

In terms of Figs. 3–5 the vicinity of the stationary state can be explored by mixing aliquots on opposite sides of the cycle; but Fig. 10 shows that unless the limit cycle is perfectly symmetric about (1, 1), the exact phasing required depends on the volume ratio $v/(1 - v)$. Figs. 6–9 are not much altered by this defection from perfect symmetry: the phaseless point is only made to bobble up and down sinusoidally as it drifts steadily to the right with increasing ϕ_2. Fig. 11 shows the ϕ_3 isochrons at each level $v =$ constant in the cube.

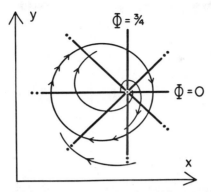

FIG. 10. This is a composite of Figs. 3 and 5, but with eccentric stationary state, retaining the circular limit cycle and equispaced radial isochrons, for simplicity. Thus the system moves slowly on the limit cycle to the right of center, but faster on the left.

Remember that phase is always measured at the moment of mixing (and at unit intervals thereafter). From Figs. 3–5 or the associated equations (5) or (7) it is evident that two aliquots will combine in the same way whenever they are mixed: adding θ to ϕ_1 and ϕ_2 leaves b and c in (7) unaltered, and can only leave a unaltered if ϕ_3 also increases by θ. So if the resultant is ϕ_3 at (ϕ_1, ϕ_2), then at $(\phi_1 + \theta, \phi_2 + \theta)$ it is $\phi_3 + \theta$, whatever θ may be. Thus the contours of constant ϕ_3 on the (ϕ_1, ϕ_2) plane are all parallel replicas, equispaced along the main diagonal. For example at $v = \frac{1}{2}$, the resultant phase is exactly intermediate between parent phases (except at $\phi_1 - \phi_2 = \pm \frac{1}{2}$), whenever they are actually mixed. In fact if the observable rhythm is any linear combination of x and y (therefore sinusoidal) then the (sinusoidal) superposition of the parent rhythms has the same phase as the resultant after mixing—which is another way of saying that measured from some fixed time, e.g., Greenwich midnight, the compromise phase does not depend on when the mixing is actually carried out[2]. Thus

---

[2]Thus if as in Hastings and Sweeney (1958) and Brinkmann (1967) the result of mixing two cell suspensions' roughly sinusoidal rhythms is indistinguishable in both phase and amplitude from the superposition of the unmixed controls, then either (as usually inferred) they are not interacting OR they do interact, even completely and immediately, but the oscillation in each cell is quite symmetric and recovers but slowly, if at all, toward a limit cycle.

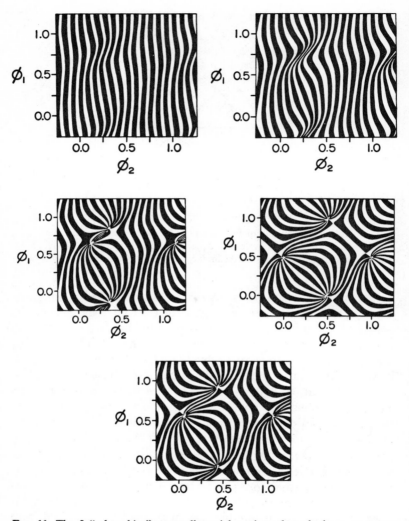

FIG. 11. The 5 "zebra-skins" are really serial sections through the $\phi_1 \times \phi_2 \times v$ cube at $v = 0.1$ (or 0.9, by interchanging subscripts 1 and 2, according to assumption B), $v = 0.2$ (or 0.8), $v = 0.3$ (or 0.7), $v = 0.4$ (or 0.6) and $v = 0.5$. The edges of the black and white stripes are loci of fixed ϕ_3. Each stripe is $1/24$ cycle wide. These contour maps were computed digitally using equations 4 and 8, so the phaseless locus is a sinusoid twice impaling plane $v = 0.5$, once in each triangle of Figs. 1 and 2. (The printout scale was about 8% greater horizontally than vertically, so the sections are not quite square; each covers $1\frac{1}{2}$ cycles along both axes in order to make plainer the isochrons' continuity on the $\phi_1 \times \phi_2$ torus except at the singular points.)

280

it would also be the same if complete mixing occurs, but slowly, as reactants diffuse through a resisting membrane or medium.

But less symmetric models, e.g., equation (8), do not strictly adhere to this rule. The eccentric limit cycle of Fig. 10 gives ϕ_3 contours for symmetric mixing (Fig. 1 at $v = \frac{1}{2}$) which are *not* all parallel displacements of a common contour along the diagonal, $\phi_1 - \phi_2 = $ constant.

GLYCOLYSIS AS A LIMIT CYCLE OSCILLATOR

Ghosh, Chance, and Pye (1971) reported the results of such experiments, using suspensions of anaerobically-metabolizing yeast cells in which glucose is passing through the Embden-Meyerhof pathway in pulses 30 seconds apart due to a limit cycle in the regulatory kinetics (Selkov, 1972; Winfree, 1972). This spontaneous periodic activity is monitored by the blue fluorescence of the crucial cofactor NADH, while the cell population is exposed to ultra-violet light. The 10^9 cells in each population apparently are metabolically coupled with sufficient strength to permit treating them collectively as a single glycolytic oscillator. Because the plasma membrane restricts exchange of phosphorylated intermediates, these interactions are apparently mediated by inorganic ions and/or pyruvate, acetaldehyde, and ethanol. Ghosh *et al.* ascribe little role to the last three.

Ghosh *et al.* mixed yeast cell suspensions of equal volume ($v = \frac{1}{2}$) at diverse combinations of phase during the NADH cycle. Soon after mixing, NADH fluorescence amplitude appears normal, suggesting strong interactions and immediate synchronization of the initially phase-separated limit-cycle oscillations. The resulting phase compromise was read from their circle diagrams and plotted in Fig. 12 on the $\phi_1 \times \phi_2$ plane at $v = \frac{1}{2}$, providing a detailed empirical substitute for "R" in each triangle of Fig. 1. These data are not incompatible with Fig. 11: only 4 out of 47 differ by more than 1/6 cycle from the contour lines sketched on the same coordinates in Fig. 13. Note that in Fig. 13 and very nearly in Fig. 12, if either aliquot is in the third quadrant ($3/4 < \phi < 1$), then so is ϕ_3. Since the results of near-replicate experiments frequently differ markedly, it is hard to imagine any smooth function $\phi_3(\phi_1, \phi_2, \frac{1}{2})$ fitting much better; but as the authors emphasize, "Much more experimentation is probably needed" before any pattern emerges unambiguously. A possible alternative is suggested by the fact that in all but 2 out of 47 experiments (and in all $\phi_1 = \phi_2$ controls, presumably), ϕ_3 differs by less than 1 on our scale of 0 to 9, from one or the other of the two parent phases. Could it be that ϕ_3 is *not* a single-valued function of ϕ_1, ϕ_2, and v, but rather may be *either* ϕ_1 or ϕ_2, depending perhaps on details of the mixing procedure? Perhaps cells initially in the minority simply adopt majority phase during injection of one aliquot into the other. If so, then ϕ_3 is not a single-valued map into S, and no discontinuity is implicit in (A), (B), (C) above.

However if there *is* a unique compromise phase, obscured only by random noise in the data of Fig. 12, then the somewhat

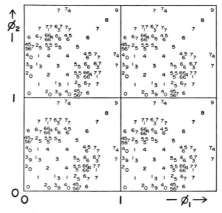

FIG. 12. Data from the 47 experiments (and implicit $\phi_1 = \phi_2 = \phi_3$ controls) of Ghosh *et al.* (1971), using glycolytic oscillations in yeast cell suspensions of equal volume. The compromise phase is plotted in the format of Figs. 1, 2, and 11. Each phase axis is quantized into 10 equal parts, and in each of the 100 squares so defined is written the resulting ϕ_3 as a digit 0–9. Digit D indicates that the result was within 1/20 cycle of D/10 cycles past NADH maximum an integer number of cycles after mixing. On this scale, NADH maximum is at $\phi = 0.05$, and minimum is $\phi = 0.55$. Some squares contain as many as 5 experiments, but most have none. Each experiment is recorded twice in each unit cell, i.e. with labels 1, 2 assigned both ways.

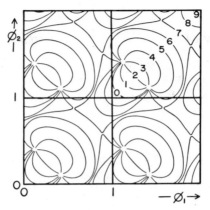

FIG. 13. ϕ_3 contour lines similar to Fig. 11 at $v = \frac{1}{2}$ are suggested as an approximation to the glycolysis data of Fig. 12.

Procrustean effort of Fig. 13 may help direct further measurements into the more revealing regions of the cube. In particular it suggests *avoiding* combination of the maximum and minimum of NADH fluorescence, where data variance is likely to be high, until the isochrons are more clearly defined by measurement in smoother regions—some of which remain completely unsampled.

THE MITOTIC CYCLE AS A RELAXATION OSCILLATOR

The phaseless locus has a quite different appearance in experiments using a simple relaxation oscillator: it is exhibited when the physiology of either oscillator undergoes its periodic jump. For example: suppose with Rusch *et al.* (1966), Cohen (1971), and Sachsenmaier *et al.* (1972) that a mitogen, *m*, accumulates in the cytoplasm (or on nuclear sites) until, reaching $m = 1$, it triggers mitosis and is destroyed (or the number of titrating nuclear sites doubles) thus initiating the next cycle.

Mixing aliquots m_1 and m_2 gives $m_3 = vm_1 + (1 - v)m_2$. Then taking ϕ mod 1 $\propto m$, we have

$$\phi_3 = v\phi_1 \bmod 1 + (1 - v)\phi_2 \bmod 1 \qquad (9)$$

(or $\phi_3 = v\phi_1 + (1 - v)\phi_2$ on the understanding that a phase is always between 0 and 1). The contours of ϕ_3 on the $\phi_1 \times \phi_2$ plane are parallel equispaced straight lines of slope $-v/(1 - v)$: parallel to the ϕ_1 axis at $v = 0$, perpendicular to the main diagonal at $v = \frac{1}{2}$, and parallel to the ϕ_2 axis at $v = 1$. A jump of $-v$ occurs everywhere along plane $\phi_1 = 0$, where m_1 changes from 1 to 0; a jump of $v - 1$ occurs everywhere along plane $\phi_2 = 0$, where m_2 changes from 1 to 0. In cross-section at fixed ϕ_1 or ϕ_2, the ϕ_3 contours are also straight lines. Fig. 14, in the format of Figs. 1 and 11, shows R at $v = \frac{1}{2}$. Conditions (A), (B), (C) are satisfied, so a discontinuity is inescapable.

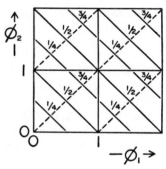

Fig. 14. As in Figs. 1, 11, and 13, ϕ_3 is portrayed by level contours in the $\phi_1 \times \phi_2$ plane at $v = \frac{1}{2}$, with the unit cell repeated twice in each direction for clarity. In this case a linear "division protein" or "mitogen" model is assumed to underlie the periodism (equation 9). Accordingly ϕ_3 is a featureless tilted plane within each unit cell. This imposes an abrupt jump at the edges, ϕ_1 or $\phi_2 = 0$. The horizontal and vertical lines are not ϕ_3 contours, but unit cell boundaries, at which ϕ_3 is discontinuous.

The result of mixing a given pair of aliquots ($\phi_1 - \phi_2 =$ constant) depends in a simple way on *when* they are fused. There are two possible results, differing by v (or by $1 - v$, which is the same thing mod 1) according to whether $\phi = 0$ lies in the larger or smaller interval between ϕ_1 and ϕ_2 at the moment of mixing. (This can be seen by adding any time t to both t_1 and t_2 in equations (9).) Sachsenmaier *et al.* (1972) have done precisely this experiment using *Physarum* with $v = \frac{1}{2}$, and in fact obtain a phase jump of $\frac{1}{2}$ cycle when plasmodial fusion begins 1 to 2 hours before mitotic telophase. Apparently then, the physiological jump in this relaxation oscillator precedes mitosis by about an hour. Note that this *form* of discontinuity is a consequence of the particular assumed mechanism; but that there *is* a discontinuity is more nearly a logical than a biological fact.

Fig. 15 places data from fusion of equal plasmodia on the unit square, together with selected isochrons $\phi_3 = R(\phi_1, \phi_2, \frac{1}{2})$ from Fig. 14. The results lack $\phi_1 = \phi_2 = \phi_3$ controls, and (due to the short duration of each experiment) are not demonstrably periodic (as condition (A) requires) after fusion of plasmodia. Nonetheless, agreement with the mitogen model seems satisfactory. The few available data at $v \neq \frac{1}{2}$ are compatible with the same interpretation. In contrast, this pattern distinctly does not approximate the data from glycolysis, in which the isochrons at $v = \frac{1}{2}$, to the extent that they can be discerned, run at right angles to the main diagonal only when close to it, but run horizontally or vertically elsewhere, so that the discontinuity is confined to two *points*.

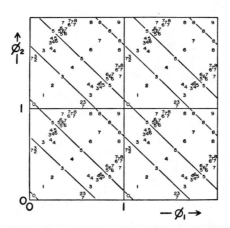

FIG. 15. The data of Rusch *et al.* (1966) and Sachsenmaier *et al.* (1972) from fusion of equal *Physarum* plasmodia are depicted in the notation of Fig. 12. A time 1.5 hours before mitotic telophase serves as phase 0, then the cycle is divided into 10 equal parts. The end of mitosis thus occurs at $\phi = .15 + \frac{1}{2} 1/10 = 0.2$ on this scale. ϕ_3 is denoted by a digit 0–9 in the square indexed by ϕ_1 and ϕ_2. As in Fig. 12, each experiment is reported once with the earlier parent plasmodium called 1, and once with it called 2. Selected contours from Fig. 14 are superimposed.

The phase mixing behavior of a simple relaxation oscillator such as the *Physarum* mitosis model (and any other "division protein" model for the cell cycle: see Zeuthen and Williams (1969)) can be seen as a limiting case of the asymmetric limit-cycle's phase-mixing behavior. The dynamical portrait corresponding to Fig. 10, for the relaxation oscillator might be as in Fig. 16: the system traverses a straight-line trajectory in concentration space from A at $t = \varepsilon$ through B to C at $t = 1 - \varepsilon$ and then in time 2ε departs swiftly from that 1-dimensional locus to return to its beginning. Almost all the isochrons are packed into the slow arc ABC (e.g., Tyson and Light (1974)). In deforming Fig. 10 to Fig. 16, the corresponding Fig. 11 is deformed to Fig. 14: a segment of the limit cycle in Fig. 10, initially spanning a small range of isochrons, is expanded to occupy almost all of the phase axes along both coordinates in Fig. 11. The corresponding small area near the diagonal in Fig. 11 is thus expanded to occupy almost the whole square, with its parallel, equally spaced ϕ_3 contours relabelled to span almost the whole cycle in Fig. 16. All the rest of the contours, including the singularities, are squeezed onto the boundaries of the square, which thus become discontinuities.

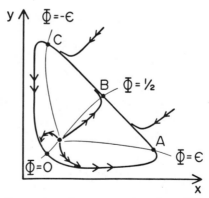

FIG. 16. A dynamical portrait on the concentration plane of Figs. 3, 4, 5, and 10, in which the limit cycle has a nearly straight arc ABC traversed at nearly constant velocity, and a curved arc traversed in a short time 2ε. Trajectories are imagined to approach the limit cycle very swiftly. This is one kind of "relaxation oscillator", possibly a suitable representation of "division protein" kinetics, in which m and ϕ in equation (9) vary almost from 0 to 1 along straight arc ABC.

DISCUSSION

Returning now to symmetry principles (A), (B), (C) and the implied discontinuity, three questions about the generality and applicability of these remarks deserve more complete answers than I have been able to provide.

First, is the *logically* necessary existence of a phaseless locus in R strictly correlated with reaching a stationary state? I think not. In dynamical systems involving more than two components, access to the stationary state cannot be guaranteed, as it need not lie in the "innards" of the cycle. But supposing the result is a single-valued map to the unit circle, a short topological argument (Appendix) indicates there must exist a "phaseless manifold" threading the limit cycle in the sense that it penetrates every cap, i.e., every surface bounded only by the limit cycle. The states accessible by chemical mixing at fixed ϕ_1 by varying ϕ_2 and v constitute such a cap (Fig. 17); accordingly at least one of them lies in the "phaseless manifold" and all phases are realized at nearby combinations of ϕ_1 and v. Thus the general appearance of Figs. 6, 7, and 11 seems to me very likely independent of the dimensionality of Figs. 3 and 10, so long as there is but one simple limit cycle. However the phaseless manifold can also have topologically more complex branches (e.g., in a trefoil-knotted limit cycle) which will appear as additional discontinuity loci in R.

Secondly, how much are the main conclusions of this analysis affected if mixing is *selective*, some reactants being restrained by a semipermeable membrane? Suppose that in the model of Figs. 3–5, mixing homogenizes the x pools, but y_1 and y_2 do not directly interact. If the coupled oscillators return to the same limit cycle as before, then the result must still have properties (A), (B), (C), and must additionally be identical to the case of unreserved mixing when $v = 0$ or 1 and when $y_1 = y_2$, i.e., in planes $\phi_1 = \phi_2$ and $\phi_1 + \phi_2 = \frac{1}{2}$, where the wheel-hub of isochrons is encountered. Is the isochron structure qualitatively different between these planes of identity to the simpler case? It is not different in the limiting cases in which every important reactant is restrained, or no important reactant is restrained. But the more interesting cases have not yet received adequate attention. By numerical simulation,

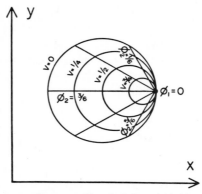

FIG. 17. On the same coordinates as Figs. 3, 4, 5, and 10 this grid shows the sets of (x, y) concentrations reached by mixing a volume v at $\phi_1 = 0$ with volume $(1 - v)$ of ϕ_2. The set accessible in this way is a 2-dimensional ruled surface bounded only by the limit cycle.

Tyson and Kauffman (1974) have found cases in which selective diffusion between two limit-cycle oscillators fails to synchronize them; i.e., ϕ_3 is not definable.

Finally, we might ask what other biological phenomena, possibly very different from mixing of oscillators, exhibit the same mathematical structure? The logic underlying the discontinuity has more to do with homotopy than with biochemistry, as G. Mitchison first pointed out to me: The problem is only that there exists no continuous map from a disk (e.g., one of the "R" triangles of Fig. 1) onto the phase circle, for which the restriction to the disk's boundary has non-zero winding number. There must exist a "phaseless" point or locus in each such disk. This dilemma has amusing consequences for other biological systems whose states can be mapped onto circles, illustrating the subtle wonders of circular logic (Winfree, 1973).

For example, one consequence is the existence of GREY. Human visual psychology is so constructed that the spectral colors map psychophysically onto a circle, the "color wheel": red is subjectively close to orange is close to yellow is close to green is close to blue is close to indigo is close to red (Sheppard,1968). By combining two colors on this circle, either in addition (with projected lights) or in multiplication (with pigments) one obtains a new color which can be identified with some spectral color in respect to hue (though not also in respect to saturation). Because by mixing two identical colors, one obtains the same hue, the same topological crisis emerges: at least for $v = \frac{1}{2}$ (equal intensities) there must exist a hueless combination of two spectral colors. It is called GREY and the component colors are called complementary.

In more quantitative terms the combination of colors is computed using Newton's barycentric average on the standard chromaticity diagram, which has radial "isochromes" (for which pun I thank S. Kauffman) as in Fig. 5 (Sheppard (1968) pp. 27–35). Taking colors 1 and 2 from any circle enclosing GREY, the resulting hue would be computed by formula (5) above. The isochromes in the $\phi_1 \times \phi_2 \times v$ cube are therefore given by Fig. 7, each ϕ_3 contour line being painted a different color in order around the circle. The locus of convergence (the phaseless manifold) is GREY.

It seems likely that this phase singularity implicit in suitably symmetric mappings $S^1 \times S^1 \to S^1$ will be encountered by students of photoperiodism, animal navigation, and circadian rhythms. Though I cannot now cite experimental examples, their emergence seems foreshadowed by the diversity of theoretical models employed in those fields, in which a daily, tidal or seasonal phase ($\in S^1$) or compass heading ($\in S^1$) is determined jointly by two phases (e.g., of dawn, dusk, phase of the moon) or by a phase and an azimuth (e.g., of circadian clock and sun, respectively) (e.g., Aschoff, 1969; Beck, 1968; Braemer, 1960; Danilevsky, 1970).

APPENDIX

Since 1968 I have been haunted by a topological conjecture. It crucially affects the interpretation of experimental projects I have since published (Winfree (1973)). Yet I can find neither proof nor counterexample. It seems the time has come to abjure vanity, and pose the question for someone else to answer. I will try to pose it in mathematically precise terms in three parts, somewhat as one uses language from a lawyer, without complete understanding. I hope the examples will make my intent clearer.

The conjecture concerns dynamical systems with asymptotically periodic behavior. Let F_t be a flow on a C^2-smooth contractible manifold C (for example R, but not S^1). The subscript t represents time; for any $c \in C$, $F_0(c) = c$ and $F_s(F_t(c)) = F_{s+t}(c)$. Let $L \subset C$ be a closed orbit of unit period: for any $l \in L$, and any integer n, $F_{t+n}(l) = F_t(l)$. Let L be an attractor (a limit-cycle), surrounded by an attractor basin B. B is an open subset of C, that is, $L \subset B \subset C$. We assume that L has "asymptotic phase" (Hale, 1969, p. 217).

For example, let F be a pair of polynomial first-order ordinary differential equations such as:

$$\dot{x} = -2\pi y + x(1 - x^2 - y^2),$$

$$\dot{y} = 2\pi x + y(1 - x^2 - y^2),$$

for which $C = R^2$, $L =$ unit circle $x^2 + y^2 = 1$, and $B = R^2$ minus a fixed-point at the origin.

The "latent phase" of any point (state) in B can now be defined by a map $\Phi: B \to S^1 (= R/Z)$ as follows. Let a "phase map" $\phi: L \to S^1$ be a homeomorphism such that $\phi(F_t(l)) = \phi(l) + t$ (modulo 1). Then define $\Phi(b) = \phi(l)$, for all points $\{b \in B: \lim_{t \to \infty} |F_t(b) - F_t(l)| = 0\}$. The set of such points (for any fixed l) is what I have been calling an "isochron", i.e., a set of initial conditions in L's attractor basin such that the system will be found at phase $t + \phi(l)$ on the limit cycle after a long time t.

The latent phase map Φ has some of the properties of ϕ, e.g., $\Phi(F_t(b)) = \Phi(b) + t$ mod 1, so $d\Phi/dt = 1$. It follows that grad Φ can vanish only where the flow is unbounded. That doesn't happen in real biochemical systems. So Φ has a non-zero gradient at every state in B.

Conjecture A: Each isochron is a smooth $(N - 1)$-dimensional submanifold of B, closed in B and diffeomorphic to R^{N-1}. The isochrons of points in L foliate B. M. Hirsch assures me that this is provable (pers. comm.), my "isochrons" being known to topologists as the "stable manifolds" of points on L. J. Guckenheimer drew my attention to a proof in theorem 3.6 of Irwin (1970).

Conjecture B: Let us define $C - B = P$, the "phaseless" manifold(s). Φ maps from B to S^1, therefore B must also be non-contractible. Since C *is* contractible, P is not empty. It can be argued that P has codimension 2, 1, or 0: let K be a "cap" on cycle L, i.e., any 2-dimensional surface bounded only by L. From homotopy theory we know there is no map from such a surface to the circle which, restricted to L, is the identity. Thus every K contains at least a point of P. In this sense, part of P "threads" L. Since the codimension of this piece of P cannot exceed the dimension of K, it must be 0, 1, or 2. Familiar differential equations provide examples of each. M. Hirsch informs me that "Antoine's necklace" provides a pathological counterexample, but that the Alexander duality theorem guarantees dim $P \geqslant N - 2$ anyway.

Conjecture C: Let δ denote the boundary of P. δ has codimension 1 or 2. Like P, it consists of trajectories which do not lead to L, and $F_t\delta = \delta$. With suitable assumptions about genericity of F, every isochron enters the neighborhood of every point of δ.

For example, in the differential equation given above, the isochrons are radii which converge to the point P thus satisfying Conjecture C. Or augment that example by a third dynamical equation $\dot{z} = -z$. Then $C = R^3$, $L =$ unit circle in plane $z = 0$, $P = \delta = z$ axis, which consists of two trajectories to the origin, and the isochrons are radial planes $y/x =$ constant, z arbitrary, all of which come arbitrarily close to every point of δ.

Alternatively, describe a dynamical system in plane *polar* coordinates by

$$\dot{R} = R(1 - R)\left(R - \tfrac{1}{2}\right),$$
$$\dot{\phi} = \pi(1 + R).$$

This has a fixed point at the origin with attractor basin extending to the anti-limit-cycle $R = \tfrac{1}{2}$, beyond which lies basin B for limit cycle $R = 1$ with period 1. P is the disk $R \leqslant \tfrac{1}{2}$, δ is the circle $R = \tfrac{1}{2}$, and the isochrons are the spirals:

$$\Phi = \phi + \ln\left(2 - 1/R\right)$$
$$\left(\Phi \text{ is undefined for } R < \tfrac{1}{2}\right)$$

each of which comes arbitrarily close to all points of δ.

A. T. Winfree

Examples of the application of these principles to various biological oscillations will be found in Winfree (1973) and earlier papers cited there, in Kauffman (1974), Kauffman and Wille (1975), Tyson and Kauffman (1974), and Pavlidis (1973).

Acknowledgements. The University of Chicago Committee on Mathematical Biology supported development of these equations and contour maps in 1969. Their comparison with recent experimental work was encouraged by stimulating exchanges with G. Mitchison in 1971, and with S. Kauffman and J. Tyson in 1972/73. Special thanks are due to G. Mitchison for persistence in trying to teach me topology. Jay Mittenthal made valuable corrections to the manuscript. I am grateful to the N. S. F. for support since 1970 (GB16513 and GB37947) and to the Public Health Service for a Research Career Development Award.

REFERENCES

1. J. Aschoff, "Phasing of diurnal rhythms as a function of season and latitude" (in German), *Oecologia*, **3** (1969), 125–165.

2. S. D. Beck, *Insect Photoperiodism*, Academic Press, New York, 1968.

3. W. Braemer, "A critical review of the sun-azimuth hypotheses," *Cold Spring Harbor Symposia on Quantitative Biology*, **25** (1960), 413–427.

4. K. Brinkmann, "Temperatureinflüsse auf die Circadiane Rhythmik von *Euglena gracilis* bei Mixotrophie und Auxotrophie," *Planta*, **70** (1966), 344–389.

5. ———, "Der Einfluss von Populationseffekten auf die Circadiane Rhythmik von *Euglena gracilis*," *Biologische Rhythmen*, **10** (1967), 1380–1400.

6. M. H. Cohen, "Models for the control of development," Symposia of the Society of Experimental Biology, **25**: *Control Mechanism of Growth and Differentiation*, Cambridge, 1971, 445–476.

7. A. S. Danilevsky, N. I. Goryshin, and V. P. Tyschenko, "Biological rhythms in arthropods," *Ann. Rev. Entomol.*, **51** (1970), 201–244.

8. W. C. Dewey, H. H. Miller, and H. Nagasawa, "Interactions between S and G 1 cells: effects on decay of synchrony," *Exp. Cell Res.*, **77** (1973), 73–78.

9. L. N. Edmunds, L. Chuang, R. M. Jarrett, and O. W. Terry, "Long-term persistence of free-running circadian rhythms of cell division in *Euglena* and the implication of autosynchrony," *J. Interdiscipl. Cycle Res.*, **2** (1971), 121–132.

10. R. Fitzhugh, "Impulses and physiological states in theoretical models of nerve membrane," *Biophys. J.*, **1** (1961), 445–566.

11. A. K. Ghosh, B. Chance, and E. K. Pye, "Metabolic coupling and synchroniza-

tion of NADH oscillations in yeast cell populations," *Arch. Biochem. Biophys.*, **145** (1971), 319–331.

12. E. Guttes, V. R. Devi, and S. Guttes, "Synchronization of mitosis in *Physarum polycephalum* by coalescence of postmitotic and premitotic plasmodial fragments," *Experientia*, **25** (1969), 615–616.

13. J. K. Hale, *Ordinary Differential Equations*, Wiley-Interscience, New York, 1969.

14. P. Hanusse, "De l'existence d'un cycle limite dans l'évolution des systemes chimiques ouverts," *C. R. Acad. Sci. Ser. C*, **274** (1972), 1245–1247.

15. J. W. Hastings and B. M. Sweeney, "A persistent diurnal rhythm of luminescence in Gonyaulax polyedra," *Biol. Bull.*, **115** (1958), 440–458.

16. M. C. Irwin, "A classification of elementary cycles," *Topology*, **9** (1970), 35–47.

17. S. Kauffman, "Measuring a mitotic oscillator: the arc discontinuity," *Bull. Math. Biol.*, **36** (1974), 171–182.

18. S. Kauffman and J. J. Wille, "The mitotic oscillator in *Physarum polycephalum*," *J. Theor. Biol.*, **55** (1975), 47–93.

19. T. Pavlidis, *Biological Oscillators: Their Mathematical Analysis*, Academic Press, New York, 1973.

20. P. N. Rao and R. T. Johnson, "Mammalian cell fusion: Studies on the regulation of DNA synthesis and mitosis," *Nature*, **225** (1970), 159–164.

21. H. P. Rusch, W. Sachsenmaier, K. Behrens, and V. Gruter, "Synchronization of mitosis by the fusion of the plasmodia of *Physarum polycephalum*," *J. Cell Biol.*, **31** (1966), 204–209.

22. W. Sachsenmaier, U. Remy, and R. Plattner-Schoble, "Initiation of synchronous mitosis in *Physarum polycephalum*," *Exp. Cell Res.*, **73** (1972), 41–48.

23. E. E. Selkov, "Nonlinear theory of regulation of the key step of glycolysis," *Studia Biophysica*, **33** (1972), 167–176.

24. J. J. Sheppard, *Human Color Perception*, Elsevier, New York, 1968.

25. J. J. Tyson and J. C. Light, "Properties of two-component bimolecular and trimolecular chemical reaction systems," *J. Chem. Phys.*, **59** (1973), 4164–4173.

26. J. J. Tyson and S. Kauffman, "Control of mitosis by a continuous biochemical oscillation: synchronization; spatially inhomogeneous oscillations," *J. Math. Biol.*, **1** (1974), 289–310.

27. A. T. Winfree, "Biological rhythms and the behavior of populations of coupled oscillators," *J. Theor. Biol.*, **16** (1967), 15–42.

28. ——, "The temporal morphology of a biological clock," in *Lectures on Mathematics in the Life Sciences* (M. Gerstenhaber, ed.), Amer. Math. Soc., Providence, **2** (1970), 109–150.

29. ———, "Oscillatory glycolysis in yeast: The pattern of phase resetting by oxygen," *Arch. Biochem. Biophys.*, **149** (1972), 388–401.

30. ———, "Time and timelessness in biological clocks," *Temporal Aspects of Therapeutics* (J. Urquardt, F. E. Yates, eds.), Plenum, New York, 1973, 35–57.

31. E. C. Zeeman, "Differential equations for the heartbeat and nerve impulse," *Towards a Theoretical Biology* (C. H. Waddington, ed.), Aldine, Chicago, **4** (1972), 8–67.

32. E. Zeuthen and N. E. Williams, "Division-limiting morphogenetic processes in *Tetrahymena*," in *Nucleic Acid Metabolism, Cell Differentiation and Cancer Growth* (E. V. Cowdry and S. Seno, eds.), Pergamon Press, Oxford, 1969, 203–216.

ISOCHRONS AND PHASELESS SETS

J. Guckenheimer

This paper deals with questions raised by Winfree [7] regarding the behavior of "isochrons" in mathematical models of biological oscillations. These mathematical questions lie within the domain of dynamical systems [5], the qualitative study of ordinary differential equations. Our discussion will ignore the biological context of these questions, but we shall attempt to present our results in as non-technical a fashion as possible. It is our intention that the exposition be both accessible to non-specialists of dynamical systems and accurate insofar as is possible. Proofs of theorems are contained in the appendices.

We begin with a description of the setting for Winfree's question. The object of ultimate interest is a biological "clock" or oscillation. A model is constructed for the oscillation based on the assumption that its dynamics are determined by the values of a finite number of physical and chemical parameters (temperatures, pressures, free energies, velocities, chemical concentrations, etc.). A multi-dimensional space M is constructed representing the possible values of all these physical and chemical quantities. To say that the dynamics of the system are determined by the values in M at any one time means that there is a *flow* $\Phi : M \times \mathbf{R} \to M$ defined by the condition that $\Phi(x, t) = y$ if the state x becomes the state y after t units of time. The map Φ is to satisfy the usual flow properties

$$\Phi(x, 0) = x \qquad (1)$$

and

$$\Phi(x, t_1 + t_2) = \Phi(\Phi(x, t_1), t_2). \qquad (2)$$

One makes the additional assumption about the model that M is a smooth manifold (usually a domain in a Euclidean space) and that Φ is a smooth map. The flow Φ then determines a *vector field* X by the equation $X(x) = \dfrac{\partial \Phi}{\partial t}(x, 0)$. Conversely, X determines Φ. It is convenient to speak sometimes of the flow Φ and sometimes of the vector field X.

Winfree assumes that Φ has a *stable limit cycle*. The *orbit* of the flow Φ through $x \in M$ is the set $\{\Phi(x, t)|t \in \mathbf{R}\}$. The orbit through x is *periodic* of period $\tau > 0$ if τ is the smallest positive number with the property that $\Phi(x, \tau) = x$. A periodic orbit γ is a *stable limit cycle* if there is a neighborhood U of γ with the property that if $y \in U$, then $d(\Phi(y, t), \gamma) \to 0$ as $t \to \infty$. The distance d here is the distance function for some metric on M.

A simple example of a vector field whose flow possesses a stable limit cycle is given by the differential equations

$$\dot{x} = -y + x(1 - x^2 - y^2)$$
$$\dot{y} = x + y(1 - x^2 - y^2)$$

in the plane $\mathbf{R}^2$. In polar coordinates these differential equations become

$$\dot{\theta} = 1$$
$$\dot{r} = r - r^3.$$

The circle $r = 1$ is a periodic orbit of period 2π. It is stable because $\dot{r} > 0$ inside the circle (except at the origin) and $\dot{r} < 0$ outside the circle. This implies that r is a monotone function on each orbit and that all orbits except the origin tend to the limit cycle $r = 1$.

If $y \in M$, x is on a stable limit cycle γ, and $d(\Phi(x, t), \Phi(y, t))$ $\rightarrow 0$ as $t \rightarrow \infty$, then the eventual behavior of the points x and y looks almost the same. Winfree describes this situation by saying that y is on the *isochron* of x. If some event occurs at one place along γ, then that event will eventually occur at the same times along the orbit starting at y as they will on the orbit starting at x. Winfree asks:

QUESTION A: *Do isochrons exist? Is a neighborhood of a stable limit cycle partitioned into the isochrons of points on the limit cycle?*

The answer to this question is yes if one places a nondegeneracy assumption on the behavior of the flow near the limit cycle. The existence of isochrons in this case is a theorem of dynamical systems which has been known for a few years. We describe now the non-degeneracy condition and state a mathematical theorem answering Question A. A proof of the theorem is discussed in Appendix A.

Let $\phi : M \times \mathbf{R} \rightarrow M$ be a flow with a periodic orbit γ of period τ and let $x \in \gamma$. A *cross-section* of γ at x is a submanifold $N \subset M$ with the following properties:

(1) $x \in N$ and $\overline{N} \cap \gamma = \{x\}$. $\overline{N}$ is the closure of N in M.
(2) $T_x N + T_x \gamma = T_x M$.

The second condition says that N is transverse to γ at x. The *Poincaré map* Θ is a map defined on a neighborhood V of x in N with image in N. The map Θ is characterized by the condition that if $y \in N$, then $\Theta(y)$ is the first point of intersection of the forward orbit of y with N when this makes sense. Since the flow Φ is continuous, Θ will be well defined in a neighborhood of x in N. The time of the first intersection will be near τ for points near x. See Fig. 1. One then says that γ is an *elementary* (or *hyperbolic*) limit cycle if the matrix $D\Theta_x$ of first partial derivatives of Θ at x has no eigenvalue of absolute value one. The eigenvalues of $D\Theta_x$ are often called the *characteristic multipliers* of γ. They are independent of the choices of x and N. If γ is an elementary, stable limit cycle, then all of its characteristic multipliers have absolute value smaller than one. Every orbit in a neighborhood of γ tends toward γ exponentially fast.

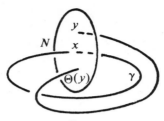

FIG. 1.

The existence of isochrons is equivalent to the existence of cross-sections to γ for which the time of first return is identically the period of γ. We seek a cross-section N for which $\Phi(N,\tau) \subset N$. If γ is a stable limit cycle, such a cross-section will be the isochron of its intersection with γ.

The concept of an isochron is closely related to the concept of a stable manifold in dynamical systems. If Φ is a flow on M and S is a subset of M, then the *stable set* of S, denoted $W^s(S)$, is the set of points y for which $d(\Phi(y, t), \Phi(S, t)) \to 0$ as $t \to \infty$. The *unstable set* of S, denoted $W^u(S)$, is $\{y \mid d(\Phi(y, t), \Phi(S, t)) \to 0$ as $t \to -\infty\}$. If an (un)stable set is also a manifold, it is called an *(un)stable manifold*. The basic theorem we state regarding Question A is the following special case of the Invariant Manifold Theorem [2, 3]:

THEOREM A: *Let $\Phi: M \times \mathbf{R} \to M$ be a smooth flow with an elementary, stable limit cycle γ. The stable set $W^s(x)$ of each $x \in \gamma$ is*

(1) *a cross-section of γ,*

(2) *a manifold diffeomorphic to Euclidean space.*
Moreover, the union of the stable manifolds $W^s(x)$, $x \in \gamma$, is an open neighborhood of γ and the stable manifold of γ.

This theorem proves the existence of isochrons for the elementary stable limit cycle γ of Φ. Note that for any flow Φ, $W^s(\Phi(x, t)) = \Phi(W^s(x), t)$. It follows from this observation that isochrons are permuted by the flow. In dealing with the second and third questions of Winfree, we shall assume that the limit cycle in question is elementary.

The second question Winfree asks is a topological question about the stable manifold of a stable limit cycle.

QUESTION B: *Suppose* $\Phi:\mathbf{R}^n \times \mathbf{R} \to \mathbf{R}^n$ *is a flow with an elementary, stable limit cycle* γ. *Does the frontier of* $W^s(\gamma)$ *have dimension* $\geq n - 2$? *The* frontier *of a set* S *is* $\bar{S} - \text{int}S$ *with* $\bar{S}$ *the closure of* S, *and* intS *the interior of* S.

M. Hirsch pointed out that the answer to the question is yes for topological reasons. The space $W^s(\gamma)$ is an open set of $\mathbf{R}^n$ homeomorphic to $S^1 \times \mathbf{R}^{n-1}$. Its first homology group is non-zero since γ cannot be deformed to a point in $W^s(\gamma)$. From the Alexander Duality Theorem [6] of algebraic topology, it follows that the complement of $W^s(\gamma)$ has a non-trivial homology class of dimension $n - 2$ that links γ. The existence of this homology class implies that the frontier of $W^s(\gamma)$ has dimension at least $n - 2$. Details and specific references are given in Appendix B.

We remark that the answer to question B is yes for a flow on a manifold M of dimension n whose homology group of dimension $n - 1$ is zero. It is not true on any manifold M as the following example demonstrates. On the three sphere S^3, let X be a vector field which points "down" except at the north and south poles where X has an elementary singular point. Let Y be a vector field on the circle S^1 which is never zero. The sum $X + Y$ defines a vector field on $S^3 \times S^1$ with two periodic orbits γ_1, γ_2 at the {north pole} $\times S^1$ and {south pole} $\times S^1$ respectively. The periodic orbit γ_2 is an elementary stable limit cycle. Moreover, $W^s(\gamma_2) = S^3 \times S^1 - \gamma_1$. Therefore the frontier of $W^s(\gamma_2)$ is γ_1 and the dimension of γ_1 is $1 < 4 - 2$.

The third question of Winfree concerns the behavior of the isochrons of a stable limit cycle γ near the frontier of $W^s(\gamma)$.

QUESTION C: *For "generic" flows* $\Phi:M \times \mathbf{R} \to M$ *possessing an elementary stable limit cycle* γ, *is it true that every neighborhood of every point on the frontier of* $W^s(\gamma)$ *intersects every isochron of* γ?

Questions A, B, and C motivate a number of experiments performed by Winfree. By adjusting a pair of experimental parameters, he is able to create experiments for which the initial conditions in a model lie on or near the frontier of $W^s(\gamma)$, γ a stable limit cycle. The experimental results for these values of the experimental parameters display one of two phenomena: (1) a destruction of the oscillation entirely, or (2) points arbitrarily close to one another lying on isochrons of every point of γ. The second possibility indicates that all of the isochrons of γ are passing arbitrarily close to a single point on the frontier of $W^s(\gamma)$. Since this situation is the one typically encountered in experiments, is it the one which is also typical of the models in an appropriate sense? This is a natural condition that a reasonable mathematical model should satisfy. The answer to Question C does not seem to exist in the literature of dynamical systems. Here we attempt an answer which appears almost satisfactory.

We illustrate the kinds of phenomena relevant to Question C which may occur for a flow by means of a few examples. The first example is a flow Φ on the plane $\mathbf{R}^2$ whose vector field in polar coordinates is given by the differential equations

$$\dot{\theta} = 1$$

$$\dot{r} = r(r^2 - r_1^2)(r^2 - r_2^2).$$

This vector field has two periodic orbits γ_1, γ_2 given by $r = r_1$ and $r = r_2$ respectively. The periodic orbit γ_1 is an elementary stable limit cycle. The stable manifold $W^s(\gamma_1)$ of γ_1, is the set $\{(r, \theta)|0 < r < r_2\}$. If $(r_1, \theta_1) = x \in \gamma$, $W^s(x)$ is easily seen to be $\{(r, \theta)|0 < r < r_2$ and $\theta = \theta_1\}$ since the angular velocity of the flow is identically one. Thus Φ is a flow which has an elementary stable limit cycle but does not satisfy the conditions specified in Question C. The frontier of $W^s(\gamma_1)$ contains γ_2. If $x \in \gamma_2$, then a small neighborhood of x does not meet each isochron of γ_1. The isochrons of γ_1 are contained in radial lines of the flow. See Fig. 2.

The second example is obtained from the first by changing the parametrization of the flow in the first example without changing

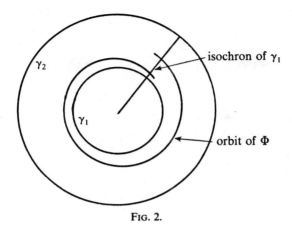

the orbits. Let $f : \mathbf{R} \to \mathbf{R}$ be a smooth function with the properties that

(1) $f(r) \geqslant 1$,
(2) $f(r) = 1$ unless r is near r_1. In particular $f(r_2) = 1$,
(3) $f(r_1) > 1$.

Let Y be the vector field $f(r)X$ with X the vector field of the first example. The differential equations defining Y are

$$\dot{\theta} = f(r)$$

$$\dot{r} = f(r)r(r^2 - r_1^2)(r^2 - r_2^2).$$

The orbits of Y are the same as those of X. There are still periodic orbits of Y at $\gamma_2 = \{(r, \theta) | r = r_2\}$; $i = 1, 2$. Unlike the first example, the periods of γ_1 and γ_2 are different for the flow of Y. The periodic orbit γ_1 has a period $2\pi / f(r_1)$, and the period of γ_2 is still 2π. The effect of this perturbation is to bend the isochrons of Y. See Fig. 3.

We are interested in whether the isochrons of Y satisfy the criteria of Question C. The key observation to be made is the following. If we follow the flow for time $2\pi / f(r_1)$ each isochron is

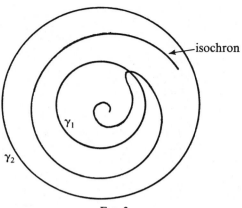

isochron

FIG. 3.

mapped into itself, while if we follow the flow for time 2π, each point of γ_2 on the frontier returns to itself. These two requirements are compatible only if the isochrons of γ_1 wind around the annulus $r_1 < r < r_2$ infinitely often as they approach γ_2. In Appendix C we give the topological arguments which prove that Y satisfies the criteria of Question C.

The first two examples considered above deal with vector fields in the plane. It is well known that there is a drastic difference in the qualitative behavior vector fields can have in two and higher dimensions. It is reasonable to expect that the discussion of Question C will be considerably more difficult in more than two dimensions. The third example is intended to demonstrate some of the complications which arise in three dimensions.

The third example is described by a pair of perturbations to one flow. Consider the flow Φ on $\mathbf{R}^2 \times S^1$ whose vector field X is defined by the differential equations

$$\dot{r} = r(1 - r^2)$$

$$\dot{\theta} = 0$$

$$\dot{\phi} = 1$$

with (r, θ) polar coordinates on $\mathbf{R}^2$ and ϕ the coordinate on S^1.

The flow Φ has an elementary stable limit cycle $\gamma = \{0\} \times S^1$. The stable manifold $W^s(\gamma) = \{(r, \Theta, \Phi)|r < 1\}$. As in the first example, the isochrons are easily described as the sets $\phi =$ constant in $W^s(\gamma)$. Denote the frontier of $W^s(\gamma)$ by B.

We now make two perturbations in X. The first perturbation X_1 is obtained by adding a small vector field to X which has a component only in the θ direction and produces a vector field X_1 such that X_1 restricted to B has two periodic orbits γ_1 and γ_2 given by the sets $\{(1, \frac{\pi}{2}, \phi)|\phi \in S^1\}$ and $\{(1, -\frac{\pi}{2}, \phi)|\phi \in S^1\}$ respectively. We assume that all other orbits of the flow of X_1 in B flow from γ_1 to γ_2. The second perturbation X_2 of X is obtained by multiplying X_1 by a function $f : \mathbf{R}^2 \times S^1 \to \mathbf{R}$ which is identically 1 outside a small neighborhood U of y_1 and which is identically $1 + \varepsilon$ inside a smaller neighborhood of γ_1.

Does X_2 satisfy the criteria of Question C? To answer this question, it is necessary to describe the isochrons of X_2. The isochrons of X_1 are the same as the isochrons of X. If $(r, \theta, \phi) \in W^s(\gamma)$ has a forward orbit for X_2 which does not intersect U, then (r, θ, ϕ) is on the isochron of the point $(0, 0, \phi) \in \gamma$ for X_2 since the velocity of X_2 in the ϕ direction is identically one outside U. The period of γ for X_2 is 2π; therefore, the map obtained by following the flow backwards for 2π maps each isochron to itself. Since each orbit inside $W^s(\gamma)$ eventually remains outside U, by iterating the -2π map of the flow, one can determine to which isochron each point belongs.

One finds that the isochrons look like the drawing of Fig. 4. The frontier of each isochron is the closure of a curve in B which

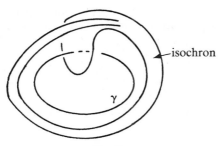

isochron

γ

FIG. 4.

approaches γ_1 from both sides. If $p \in B - \gamma_1$, then small neighborhoods of p do not intersect every isochron of γ. If $p \in \gamma_1$, then every isochron contains p in its frontier. On the "half-cylinder" $\theta = \frac{\pi}{2}$, this example looks like the second example considered above. Elsewhere it has the general features of the first example in relation to the criteria of Question C.

We now state our results regarding Question C. By reparametrization of vector fields along stable limit cycles, it is possible to prove that the set of vector fields meeting the criteria of Question C is dense in the set of all vector fields having an elementary, stable limit cycle (in the space of C^r vector fields with the C^r topology). The question of whether the set is "generic" seems to be difficult. The results we have obtained are contained in the following three theorems:

THEOREM C1: *Let M be a compact oriented two dimensional manifold. In the space of C^r vector fields on M having an elementary stable limit cycle γ, there is an open-dense set of vector fields for which every neighborhood of every point in the frontier of $W^s(\gamma)$ intersects each isochron of γ.*

THEOREM C2: *Let X be a C^r vector field on a manifold M having an elementary stable limit cycle γ. In the space of vector fields Y such that $Y = X$ on a neighborhood of the complement of $W^s(\gamma)$, there is a neighborhood $\mathcal{U}$ of X and a generic subset $\mathcal{C}$ of $\mathcal{U}$ such that every vector field in $\mathcal{C}$ has a limit cycle β with the property that every neighborhood of every point in the frontier of $W^s(\beta)$ intersects each isochron of β. "Generic" here is used in the sense that a subset of a topological space is generic if it is a countable intersection of open-dense sets.*

THEOREM C3: *Let M be a compact manifold and let Σ be the space of vector fields on M satisfying Smale's Axiom A', the strong Transversality Property, and having a stable limit cycle γ. There is a dense open subset of vector fields in Σ with the property that every neighborhood of every point in the frontier of $W^s(\gamma)$ meets each isochron of γ.*

Proofs of these theorems are given in Appendix C. The difficulty in removing the restriction to a class of structurally stable vector fields in Theorem C3 lies in the fact that one has little control on the frontier of the stable manifold of a stable, elementary limit cycle. This frontier may be drastically altered by a perturbation of the vector field. There is no apparent relation between the behavior of two nearby vector fields on the frontier of a common stable limit cycle.

Appendix A

In this appendix we shall prove the existence of "isochrons" for an elementary, stable limit cycle. This result is not new [3]. Much more extensive theorems have been proved by Hirsch, Pugh, and Shub [2]. Our purpose here is to give a reasonably elementary, self-contained proof of the result needed for our applications.

It is useful to reduce the problem somewhat and cast it in slightly different terms. The initial assumption is that the flow Φ has a limit cycle γ of period τ. The map $f = \Phi(\cdot, \tau): M \to M$ has γ as a set of fixed points. Moreover, if we consider Df along γ, it has an eigenvector with eigenvalue 1 along γ. All other eigenvalues of Df along γ have absolute value less than one. This implies that we can find a neighborhood U of γ and a metric on M with the properties $f(U) \subset U$ and $d(f(x), \gamma) < d(x, \gamma)$ for all $x \in U$. We can parametrize U as a neighborhood of $\{0\} \times S^1 \subset \mathbf{R}^{n-1} \times S^1$ in such a way that at each $x \in \gamma$, Df_x leaves invariant the hyperplane of $T_x M$ tangent to $\mathbf{R}^{n-1}$ and is a contraction on this hyperplane.

PROPOSITION: *Let $f: \mathbf{R}^{n-1} \times S^1 \to \mathbf{R}^{n-1} \times S^1$ be the time τ map of a flow Φ satisfying*

(1) *$\{0\} \times S^1$ is a periodic orbit γ of Φ of period τ.*
(2) *If $x \in \gamma$, then Df_x leaves invariant the subspace tangent to $\mathbf{R}^{n-1}$ and has norm smaller than one on this subspace.*
(3) *Then there exist invariant manifolds $W(x)$, $x \in \gamma$ such that $W(x) = \{ y \in \mathbf{R}^{n-1} \times S^1 | d(f^n(y), x) \to 0$ as $n \to \infty \}$.*
(4) *The union of the $W(x)$ is a neighborhood of γ.*
(5) *The flow Φ permutes the $W(x)$.*
(6) *$W(x)$ and $\mathbf{R}^{n-1}$ have the same tangent space at x.*

The proof of the proposition uses the following lemma which uses the same notation as the proposition:

LEMMA: *The sequences of functions* $\{f^n\}$ *and* $\{Df^n\}$ *are uniformly convergent sequences of functions in a neighborhood of* γ. *The function* $\lim_{n\to\infty} Df^n$ *has constant rank* 1.

Proof: We need to establish some notation. If $z = (y, \theta) \in \mathbf{R}^{n-1} \times S^1$ then $|z|$, $|y|$, and $|\theta|$ will used for the norms of z, y, θ in suitable coordinate systems on $\mathbf{R}^{n-1} \times S^1$, $\mathbf{R}^{n-1}$, and S^1 respectively. We shall denote by π_1 and π_2 the projections of $\mathbf{R}^{n-1} \times S^1$ onto $\mathbf{R}^{n-1}$ and S^1. If $\varepsilon > 0$, there is a neighborhood U of γ such that if $z = (y, \theta) \in U$, then the following estimates hold:

(1) $|\pi_2 f(z) - \pi_2(z)| \leqslant \varepsilon|\pi_1(z)|$.
(2) $|\pi_1 f(z)| \leqslant \mu|\pi_1(z)|$ for some fixed $0 < \mu < 1$ independent of z.
(3) $|Df(z) - D\pi_2(z)| < \varepsilon|\pi_1 z|$.

The first two of these estimates easily imply that the sequence $\{f^n\}$ is uniformly convergent in U. If $z_1, z_2 \in U$, then for large enough n we have

$$|f^n(z_1) - f^n(z_2)| < \varepsilon + |\pi_2 f^n(z_1) - \pi_2 f^n(z_2)|$$

since $f^n(z) \to \gamma$ uniformly by estimate (2) above. Now estimates (1) and (2) imply

$$|\pi_2 f^n(z) - \pi_2(z)| \leqslant \sum_{i=1} |\pi_2 f^i(z) - \pi_2 f^{i-1}(z)|$$

$$\leqslant \varepsilon \sum_{i=1} \mu^{i-1}|\pi_1(z)| < \frac{\varepsilon}{1-\mu}|\pi_1(z)|.$$

Hence

$$|\pi_2 f^n(z_1) - \pi_2 f^n(z_2)| < \frac{\varepsilon}{1-\mu}\left(|\pi_1(z_1)| + |\pi_1(z_2)|\right)$$

$$+ |\pi_2(z_1) - \pi_2(z_2)|.$$

This yields an estimate for $|f^n(z_1) - f^n(z_2)|$ independent of n, proving that the sequence f^n is uniformly convergent.

To prove that Df^n is uniformly convergent, it suffices to note that $Df^n(z) = Df(f^{n-1}(z)) \circ \cdots \circ Df(f(z)) \circ Df(z)$. We have an estimate for each term of this composition as $D\pi_2 + E_i$ where $|E_i| < \varepsilon \mu^{i-1} |\pi_1 z|$. Since $D\pi_2$ is a projection $(D\pi_2 \circ D\pi_2 = D\pi_2)$, this yields an estimate for $Df^n(z)$ as $D\pi_2 + |\pi_1 z| \cdot$ bounded term. Thus Df^n is uniformly convergent.

It remains to prove that the rank of $\lim Df^m$ is identically one. Since $f^m(z) \to \gamma$ uniformly at an exponential rate, it follows from the mean value theorem that, for sufficiently large m, $n - 1$ of the eigenvalues of Df^m will be arbitrarily close to zero. Hence the kernel of $\lim_{m \to \infty} Df^m$ will have dimension $n - 1$. This finishes the proof of the lemma.

The lemma implies that the function $g = \lim_{m \to \infty} f^m$ is a submersion onto γ in a neighborhood of γ. The implicit function theorem [1] implies that the inverse image of $x \in \gamma$ is a smooth submanifold $W(x)$ transverse to γ. If $z \in W(x)$, then $f^m(z) \in W(x)$ and $f^m(z) \to x$ as $m \to \infty$. Therefore $W(x)$ is the manifold required by the proposition in a neighborhood U of γ. To find the remainder of the isochron of x inside $W^s(\gamma)$, we need merely form $\bigcup_{m \geqslant 0} f^{-m}(W(x))$. Since g is a submersion, there is a neighborhood D of x in $W(x)$ which is diffeomorphic to a disk with the property that $f(D) \subset D$. This implies that $\bigcup_{m \geqslant 0} f^{-m}(D)$ is diffeomorphic to Euclidean space since it is an increasing union of disks.

Appendix B

This appendix is devoted to the proof of the following theorem:

THEOREM B: *Let Φ be a flow on R^n having an elementary stable limit cycle γ. If W is the stable manifold of γ, then the dimension of $\overline{W} - W$ is at least $n - 2$. ($\overline{W}$ is the closure of W.)*

As M. Hirsch pointed out to me, the proof is a corollary of the Alexander Duality Theorem [6: p. 296, 6.2.16] which states that

$\tilde{H}_q(\mathbf{R}^n - A) = \overline{H}^{n-q-1}(A)$ if A is a compact subset of $\mathbf{R}^n$. The notation here is that $\tilde{H}_q$ is the reduced singular homology group of dimension q and $\overline{H}^k(A)$ is the direct limit of the cohomology groups $H^k(U)$ for neighborhoods U of A.

If $\overline{W}$ is compact, we apply the theorem with $A = \overline{W} - W$ and $q = 1$. The theorem implies $H_1(\mathbf{R}^n - (\overline{W} - W)) = \overline{H}^{n-2}(\overline{W} - W)$. Now W is a component of $\mathbf{R}^n - (\overline{W} - W)$ and $H_1(W) = H_1(\gamma) = Z$. Hence $\overline{H}^{n-2}(\overline{W} - W) \neq 0$. This implies that the Čech $(n-2)$-cohomology of $\overline{W} - W$ is non-trivial [6: p. 316, 6.6.2 and p. 334, 6.8.8]. This means that the dimension of $\overline{W} - W$ is at least $n - 2$ as was to be proved.

The case in which $\overline{W}$ is not compact is easily reduced to this one. Form the one point compactification S^n of $\mathbf{R}^n$ and apply the above argument on S^n instead of $\mathbf{R}^n$. Note that there is a flow on $\mathbf{R}^3$ with a stable limit cycle γ such that the complement of the stable manifold of γ is a line in $\mathbf{R}^3$. This line is not a homology 1-cycle of $\mathbf{R}^n$. See Fig. 5.

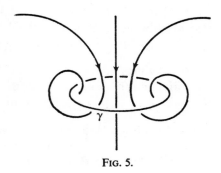

FIG. 5.

Appendix C

This appendix contains proofs of theorems providing partial answers to Question C of Winfree. These theorems are new results on dynamical systems. Throughout this appendix Φ will be a flow with vector field X on a manifold M. The flow Φ is assumed to have a stable elementary limit cycle γ of period τ with stable manifold W. The frontier $\overline{W} - W$ will be denoted B. A point $x \in B$ will be called *phaseless* if every neighborhood of x intersects the stable manifold (isochron) of each point of γ.

We shall first consider the case in which M is two dimensional and $\overline{W} \subset M$ is compact. Peixoto's Theorem [4] implies that there is an open-dense set of vector fields on M having γ as an elementary stable limit cycle which are structurally stable on a neighborhood of $\overline{W}$. If X is structurally stable in a neighborhood of $\overline{W}$, then X has a finite number of singular points on B, all of which are saddles or sources. There are no non-periodic recurrent orbits in B. We shall assume that X has these properties. For each component of B, there are two cases to consider: (1) the component of B contains a singular point of X, or (2) the component of B is a periodic orbit of X.

CASE 1: If a component β of B is a single point, it is phaseless. If β is larger than a single point and contains a singular point of X, then β contains a saddle point p of X. The unstable manifold $W^u(p)$ is an orbit of the flow. Hence $W^u(p)$ intersects each isochron transversely. Let σ be a closed curve consisting of a segment of $W^u(p)$ and a segment of an isochron I, both τ flow units apart and chosen so that σ is not contractible in the annulus A between β and γ. Under the $-\tau$ time map of the flow Φ, σ is carried onto another closed curve lying in $I \cup W^u(p)$. As the map $\Phi_{-\tau}$ is iterated, $\Phi(\sigma, -n\tau)$ tends toward β. Since σ is a cycle in the annulus A, it follows that β will be contained in the closure of $\cup \Phi(\sigma, -n\tau) \subset I \cup W^u(p)$. This implies that β is contained in the closure of I. We have proved that if β contains a saddle point, then every isochron contains β in its closure.

CASE 2: The boundary component β of B is a periodic orbit of X. We assume that A is a two dimensional annulus with boundary $\beta \cup \gamma$ and that every orbit of the flow in A goes from β to γ. Both β and γ are elementary limit cycles of X.

PROPOSITION: *If the period τ' of β is different from the period τ of γ, then each point of β is phaseless.*

Proof: Let I be the stable manifold (isochron) of $x \in \gamma$. Consider $\bar{I} \cap \beta$. This is a closed set invariant under the time τ map of the flow. We assert that $\bar{I} \cap \beta$ is also connected. If $\bar{I} \cap \beta$ can be written as the union of two disjoint, closed sets K and L and if U and V are disjoint neighborhoods of K and L, then all but a compact portion of $I \cap A$ lies in $U \cup V$. This is impossible since I keeps returning to both U and V as one approaches β on I. Therefore $\bar{I} \cap \beta$ is connected. If $\tau' \neq nt$, then the time τ map of the flow is not the identity on β. A non-trivial rotation of the circle contains no non-trivial closed connected invariant sets. Thus, if τ' is not a multiple of τ, $\bar{I} \cap \beta = \beta$. This proves the proposition unless τ' is a multiple of τ.

If τ' is a multiple of τ but $\tau \neq \tau'$, then we can apply the above argument to the flow of the vector field $-X$, interchanging the roles of β and γ. The argument then proves that the unstable manifold of each point of β for X contains γ in its closure. We conclude that the unstable manifold of each point of β intersects the stable manifold of each point of γ. This implies the proposition.

These arguments for flows on two-dimensional manifolds combine to prove the following theorem:

THEOREM C1: *On a compact, oriented two dimensional manifold M, there is an open-dense set of vector fields in the space of C^r vector fields $(1 \leqslant r \leqslant \infty)$ with the property that if γ is an elementary stable limit cycle, then every point on the frontier of the stable manifold of γ is phaseless.*

The theorem follows from the above discussion upon noting that the period of an elementary limit cycle is a continuous function on the set of structurally stable vector fields.

In order to deal with Theorem C2 on higher dimensional manifolds, it is useful to consider rather specific perturbations of a given flow. Let Φ be a flow with an elementary, stable limit cycle γ. We denote the stable manifold of γ by W and the frontier of W by B. There is a diffeomorphism $\rho: R^{n-1} \times S^1 \to W$ such that the image of $\mu(\mathbf{R}^{n-1} \times \{\theta\})$ is the stable manifold of $\mu(0, \theta)$. The map μ establishes a coordinate system on W which is well suited for our purposes. The vector field of Φ will be denoted by X.

We want to consider flows Φ' which are reparametrizations of Φ. The vector field of Φ is $fX = X'$ for some function f. We shall assume that f is a function of a particular sort.

Choose a neighborhood V of γ so that ∂V, the boundary of V, is smooth and transverse to γ and choose $\varepsilon > 0$ a small number. We require:

(1) The function f is to be identically $(1 + \varepsilon)$ on V.

(2) The function f is to be constant on each translate of ∂V under the flow Φ.

(3) There is a translate U of V so that f is identically 1 outside U.

(4) $X \cdot df > 0$ on $U - \overline{V}$.

If f is chosen in this manner and if Φ' is the flow of $f \cdot X = X'$, then the isochrons of Φ' "wrap around" W in the coordinate system described above. Our meaning is described more precisely by the next lemma.

There are two functions we wish to consider. The first ρ: $W - V \to \mathbf{R}^+$ is defined by $\rho(z) = t$ if $\Phi'(z, t) \in \partial V$. The second is the S^1 coordinate function π: $W \to S^1$ defined by identifying S^1 and γ and mapping the isochrons of W for Φ to their intersection with γ.

LEMMA: (1) *Let J be an isochron of Φ' and R a level surface of ρ. Then π is constant on $J \cap R$. Equivalently $J \cap R$ lies in an isochron of Φ.* (2) *Let α: $\mathbf{R} \to W - V$ be a curve such that* (i) *$\alpha(0) \in \partial V$,* (ii) *the image of α is contained in an isochron J of Φ', and* (iii) *$dp\left(\dfrac{d\alpha}{ds}\right) \geqslant \delta$ for some $\delta > 0$. Then $\dfrac{d\pi \circ \alpha}{ds}$ is positive and bounded away from zero in $W - U$.*

Proof: The first assertion of the lemma is a direct consequence of the definition of f. There are $t \in \mathbf{R}$ and $\theta \in S'$ such that $\partial V \cap \pi^{-1}(\theta)$ flowed backwards for time t under the flow Φ' is the set $J \cap R$. In the coordinate system determined by the map μ, the component of X' in the S^1 coordinate direction is constant on sets of the form $\Phi'(\partial V \cap \pi^{-1}(\theta), t)$. This implies that π remains constant on each set of the form $\Phi'(\partial V \cap \pi^{-1}(\theta), t)$.

Consider the map (ρ, π): $W \to \mathbf{R} \times S^1$ given by $(\rho, \pi)(z) =$

$(\rho(z), \pi(z))$. As a consequence of the first part of the lemma, $(\rho, \pi)|J$ has a smooth curve as its image. This reduces the proof of the second part of the lemma to a two dimensional problem. Outside $\bar{V}$, we assert that the image of $(\rho, \pi)|J$ is a curve of positive slope. The reason is that the component of X' in the S^1 direction is a decreasing function of ρ in $U - V$. Moreover, $\rho(z)$ represents the Φ' time it takes z to reach ∂V. Thus, in $\Phi'(V, -\tau') - V$, the larger the value of ρ, the longer it takes a point to return to another point with the same S^1 coordinate. This means that the images of isochrons of Φ' by the map (ρ, π) have positive slope outside of $\bar{V}$. The rate at which the slope changes at a point z is determined by $f(z) - (1 + \varepsilon)$.

LEMMA: *Let K be a compact neighborhood of γ in W, let I be an isochron of Φ, and let J be an isochron of Φ'. Then $I \cup J$ separates $W - K$ into at least two components. (Indeed, $I \cup J$ separates $W - K$ into a countable number of components.)*

Proof: The compact set K is contained in a set $\{z|\rho(z) < r\}$ for some $r > 0$. Thus we may assume that K is a set of this form. The image of $I \cup J$ by the map (ρ, π) is a one dimensional set which is the union of two curves. The image of I is a set of the form $\{(\rho, \theta)|\theta = \theta_0\}$ for some θ_0. The image of J is a curve along which $\rho(\theta)$ is a function with positive derivative bounded away from zero outside the image of U. It follows that the image of J has repeated intersections with I, and that the image of $(I \cup J) - K$ contains curves which separate $\mathbf{R} \times S^1 - (\rho, \pi)(K)$. Since $I \cup J = (\rho, \pi)^{-1}((\rho, \pi)(I \cup J))$. $I \cup J$ separates $W - K$.

Let us return now to the consideration of a flow Φ and a point $z \in B$ which is not phaseless. There is an isochron I of Φ such that $z \in I$. If Φ' is a perturbation of Φ obtained by reparametrization of Φ in the manner described above, then we assert that z is phaseless for Φ'. If U is a neighborhood of z and β is any curve connecting $B \cap \bar{U}$ to γ in $W - I$, then β must intersect each isochron J of Φ' outside a compact neighborhood of γ. This implies that each isochron J intersects U. This z is phaseless for Φ'.

Let us go farther. Consider the space Γ of vector fields which equal X in a neighborhood of the complement of $W = W^s(\gamma)$. We shall say a flow belongs to Γ if its vector field is in Γ. If z, Φ, and Φ' are as in the last lemma, we assert that z is phaseless for all flows belonging to Γ which are close enough to Φ'. Flows Φ'' near Φ' will satisfy the previous lemma. Isochrons of Φ'' will intersect the isochrons of Φ transversely; consequently, they will "wind around" S^1 in the coordinate system used in the lemma. The same argument which implies that z is phaseless for Φ' establishes that z is phaseless for Φ''. Moreover, note that the set of points in the frontier of W which are not phaseless for Φ form an open set. The perturbation Φ' and further perturbations Φ'' make all these points phaseless. This discussion is summarized by the following proposition:

PROPOSITION: *Let Γ be the space of vector fields which equal X in a neighborhood of the complement of $W = W^s(\gamma)$. If z in the frontier of W is not phaseless for the flow Φ of X, then there is a neighborhood U of z in the frontier of W, a perturbation Φ' of Φ belonging to Γ and a neighborhood $\mathcal{U}$ of Φ'' in the space of flows belonging to Γ such that if $z' \in U$ and $\Phi'' \in \mathcal{U}$, then z' is phaseless for Φ''.*

With this proposition, we are finally in a position to consider theorem C2.

Proof of theorem C2: Consider a one parameter family of reparametrizations of the flow Φ of the type considered in the above lemmas with the period of γ an increasing function of the parameter. Each point of the frontier of $W^s(\gamma)$ is not phaseless for at most one value of the parameter. Moreover, if z is not phaseless for the parameter value t, then there is a neighborhood of z in the frontier of $W^s(\gamma)$ which is not phaseless for the parameter value t. With the exception of at most a countable number of parameter values, the flows in the one parameter family will have the property that all points in the frontier of $W^s(\gamma)$ will be phaseless. This proves the density assertion of theorem C2. In addition, we have proved that the dense set of flows we have found are interior points of the set of flows for which all points of the frontier of $W^s(\gamma)$ are phaseless. This proves theorem C2.

We now consider theorem C3:

Proof of theorem C3: The density assertion of theorem C3 follows from theorem C2. The openness assertion will be proved by examining the periodic orbits in the frontier of $W^s(\gamma)$. First, we reduce the openness question to one involving periodic orbits. This requires a digression concerning the qualitative structure of vector fields in Σ. Unfortunately, this discussion relies much more heavily on difficult theorems from dynamical systems than the remainder of the paper. The arguments are presented in much less detail than the previous ones.

If X is a vector field in Σ with flow Φ, then there are a finite number of compact sets $\Omega_1, \ldots, \Omega_m$ with the properties:

(1) Each Ω_i is invariant under the flow Φ and contains an orbit of Φ which is dense in Ω_i.

(2) The set of periodic orbits in Ω_i is dense in Ω_i.

(3) The set of ω-limit points of each orbit is contained in one of the sets Ω_i.

(4) Each point in Ω_i has stable and unstable manifolds of complementary dimension. The union of the (un)stable manifolds of points in Ω_i is the (un)stable set of Ω_i[5].

Denote the frontier of $W^s(\gamma)$ by B. From the first property listed above one can conclude that either Ω_i and B are disjoint or $\Omega_i \subset B$. We shall ignore those Ω_i which do not intersect B or γ. (Note that γ is one of the Ω_i.) Let β_i be a periodic orbit in Ω_i if Ω_i is larger than a single point. Then the openness assertion of the theorem reduces to the following two lemmas:

LEMMA: *If each β_i is phaseless for Φ, then every point of B is phaseless for Φ.*

The vector fields of Σ are structurally stable. This implies that the Ω_i and the β_i vary continuously with perturbation. The periods of the β_i also vary continuously with perturbation. Therefore, there is an open-dense set of vector fields in Σ for which the period of each β_i is not a multiple of the period of γ.

LEMMA: *If the period of β_i is not a multiple of the period of γ for the flow Φ, then the points of β_i are phaseless for Φ.*

To prove theorem C3, it remains to prove these two lemmas. The proof of the second lemma follows the same argument as the proof of theorem C1. If I is an isochron of γ and τ is the period of γ, then $\bar{I} \cap \beta_i$ is a closed, connected subset of β_i invariant under the time τ map of Φ. If the time τ map of Φ is not the identity on β_i, then $\beta_i \subset \bar{I}$ for every isochron I. This proves the second lemma.

Only the proof of the first lemma remains. The key observation is that the intersection of $W^u(\beta_i)$ with any isochron I is always transverse. The reason is that I is transverse to the vector field X and X is tangent to $W^u(\beta_i)$. If $x \in \beta_i$ and $y \in W^s(x)$, it follows that $\overline{W^u(y)} \supset W^u(x)$. Moreover, if $I \cap W^u(x) \neq \varnothing$, then the transversality observation made above implies that $I \cap W^u(y)$. This implies $\gamma \in I$. The unstable manifolds of the β_i contain dense subsets of the Ω_i. Consequently, the union of the $W^u(\beta_i)$ contains a dense subset of B. Since the set of phaseless points is closed in B, every point of B is phaseless. This finishes the proof of the first lemma and theorem C3.

REFERENCES

1. J. Dieudonné, *Foundations of Modern Analysis*, vol. 1, Academic Press, New York, 1969.

2. M. Hirsch, C. Pugh, and M. Shub, *Invariant Manifolds*, Lecture Notes in Mathematics, No. 583, Springer-Verlag, New York, 1977.

3. M. Irwin, "A classification of elementary cycles," *Topology*, 9 (1970), 35–48.

4. M. Peixoto, "Structural stability on two dimensional manifolds," *Topology*, 1 (1962), 101–120.

5. S. Smale, "Differentiable dynamical systems ," *Bull. Amer. Math. Soc.*, 73 (1967), 747–817.

6. E. Spanier, *Algebraic Topology*, McGraw-Hill, New York, 1966.

7. A. Winfree, "Patterns of phase compromise in biological cycles," *J. Math. Biol.*, 1 (1974), 73–95.

AUTHOR INDEX

This index covers MAA Studies Volume 15 (Studies in Mathematical Biology, Part I, pages 1 to 315) and MAA Studies Volume 16 (Studies in Mathematical Biology, Part II, pages 317–624).

SUBJECT INDEX

This index covers MAA Studies Volume 15 (Studies in Mathematical Biology, Part I, pages 1 to 315) and MAA Studies Volume 16 (Studies in Mathematical Biology, Part II, pages 317–624).